Journal of Applied Logics - IfCoLog Journal of Logics and their Applications

Volume 5, Number 6

September 2018

Disclaimer

Statements of fact and opinion in the articles in Journal of Applied Logics - IfCoLog Journal of Logics and their Applications (JAL-FLAP) are those of the respective authors and contributors and not of the JAL-FLAP. Neither College Publications nor the JAL-FLAP make any representation, express or implied, in respect of the accuracy of the material in this journal and cannot accept any legal responsibility or liability for any errors or omissions that may be made. The reader should make his/her own evaluation as to the appropriateness or otherwise of any experimental technique described.

© Individual authors and College Publications 2018
All rights reserved.

ISBN 978-1-84890-288-6
ISSN (E) 2055-3714
ISSN (P) 2055-3706

College Publications
Scientific Director: Dov Gabbay
Managing Director: Jane Spurr

http://www.collegepublications.co.uk

Printed by Lightning Source, Milton Keynes, UK

All rights reserved. No part of this publication may be reproduced, stored in a retrieval system or transmitted in any form, or by any means, electronic, mechanical, photocopying, recording or otherwise without prior permission, in writing, from the publisher.

Editorial Board

Executive Editors
Dov M. Gabbay, Sarit Kraus and Jörg Siekmann

Editors

Marcello D'Agostino	Melvin Fitting	Henri Prade
Natasha Alechina	Michael Gabbay	David Pym
Sandra Alves	Murdoch Gabbay	Ruy de Queiroz
Arnon Avron	Thomas F. Gordon	Ram Ramanujam
Jan Broersen	Wesley H. Holliday	Christian Retoré
Martin Caminada	Sara Kalvala	Ulrike Sattler
Balder ten Cate	Shalom Lappin	Jane Spurr
Agata Ciabttoni	Beishui Liao	Kaile Su
Robin Cooper	David Makinson	Leon van der Torre
Luis Farinas del Cerro	George Metcalfe	Yde Venema
Esther David	Claudia Nalon	Rineke Verbrugge
Didier Dubois	Valeria de Paiva	Heinrich Wansing
PM Dung	Jeff Paris	Jef Wijsen
Amy Felty	David Pearce	John Woods
David Fernandez Duque	Brigitte Pientka	Michael Wooldridge
Jan van Eijck	Elaine Pimentel	

Area Scientific Editors

Philosophical Logic
Johan van Benthem
Lou Goble
Stefano Predelli
Gabriel Sandu

New Applied Logics
Walter Carnielli
David Makinson
Robin Milner
Heinrich Wansing

Logic and category Theory
Samson Abramsky
Joe Goguen
Martin Hyland
Jim Lambek

Proof Theory
Sam Buss
Wolfram Pohlers

Logic and Rewriting
Claude Kirchner
Jose Meseguer

Human Reasoning
Peter Bruza
Niki Pfeifer
John Woods

Modal and Temporal Logic
Carlos Areces
Melvin Fitting
Victor Marek
Mark Reynolds.
Frank Wolter
Michael Zakharyaschev

Automated Inference Systems and Model Checking
Ed Clarke
Ulrich Furbach
Hans Juergen Ohlbach
Volker Sorge
Andrei Voronkov
Toby Walsh

Formal Methods: Specification and Verification
Howard Barringer
David Basin
Dines Bjorner
Kokichi Futatsugi
Yuri Gurevich

Logic and Software Engineering
Manfred Broy
John Fitzgerald
Kung-Kiu Lau
Tom Maibaum
German Puebla

Logic and Constraint Logic Programming
Manuel Hermenegildo
Antonis Kakas
Francesca Rossi
Gert Smolka

Logic and Databases
Jan Chomicki
Enrico Franconi
Georg Gottlob
Leonid Libkin
Franz Wotawa

Logic and Physics (space time. relativity and quantum theory)
Hajnal Andreka
Kurt Engesser
Daniel Lehmann
Istvan Nemeti
Victor Pambuccian

Logic for Knowledge Representation and the Semantic Web
Franz Baader
Anthony Cohn
Pat Hayes
Ian Horrocks
Maurizio Lenzerini
Bernhard Nebel

Tactical Theorem Proving and Proof Planning
Alan Bundy
Amy Felty
Jacques Fleuriot
Dieter Hutter
Manfred Kerber
Christoph Kreitz

Logic and Algebraic Programming
Jan Bergstra
John Tucker

Logic in Mechanical and Electrical Engineering
Rudolf Kruse
Ebrahaim Mamdani

Logic and Law
Jose Carmo
Lars Lindahl
Marek Sergot

Applied Non-classical Logic
Luis Farinas del Cerro
Nicola Olivetti

Mathematical Logic
Wilfrid Hodges
Janos Makowsky

Cognitive Robotics: Actions and Causation
Gerhard Lakemeyer
Michael Thielscher

Type Theory for Theorem Proving Systems
Peter Andrews
Chris Benzmüller
Chad Brown
Dale Miller
Carsten Schlirmann

Logic Applied in Mathematics (including e-Learning Tools for Mathematics and Logic)
Bruno Buchberger
Fairouz Kamareddine
Michael Kohlhase

Logic and Computational Models of Scientific Reasoning
Lorenzo Magnani
Luis Moniz Pereira
Paul Thagard

Logic and Multi-Agent Systems
Michael Fisher
Nick Jennings
Mike Wooldridge

Logic and Neural Networks
Artur d'Avila Garcez
Steffen Holldobler
John G. Taylor

Logic and Planning
Susanne Biundo
Patrick Doherty
Henry Kautz
Paolo Traverso

Algebraic Methods in Logic
Miklos Ferenczi
Robin Hirsch
Idiko Sain

Non-monotonic Logics and Logics of Change
Jurgen Dix
Vladimir Lifschitz
Donald Nute
David Pearce

Logic and Learning
Luc de Raedt
John Lloyd
Steven Muggleton

Logic and Natural Language Processing
Wojciech Buszkowski
Hans Kamp
Marcus Kracht
Johanna Moore
Michael Moortgat
Manfred Pinkal
Hans Uszkoreit

Fuzzy Logic Uncertainty and Probability
Didier Dubois
Petr Hajek
Jeff Paris
Henri Prade
George Metcalfe
Jon Williamson

Scope and Submissions

This journal considers submission in all areas of pure and applied logic, including:

- pure logical systems
- proof theory
- constructive logic
- categorical logic
- modal and temporal logic
- model theory
- recursion theory
- type theory
- nominal theory
- nonclassical logics
- nonmonotonic logic
- numerical and uncertainty reasoning
- logic and AI
- foundations of logic programming
- belief revision
- systems of knowledge and belief
- logics and semantics of programming
- specification and verification
- agent theory
- databases

- dynamic logic
- quantum logic
- algebraic logic
- logic and cognition
- probabilistic logic
- logic and networks
- neuro-logical systems
- complexity
- argumentation theory
- logic and computation
- logic and language
- logic engineering
- knowledge-based systems
- automated reasoning
- knowledge representation
- logic in hardware and VLSI
- natural language
- concurrent computation
- planning

This journal will also consider papers on the application of logic in other subject areas: philosophy, cognitive science, physics etc. provided they have some formal content.

Submissions should be sent to Jane Spurr (jane.spurr@kcl.ac.uk) as a pdf file, preferably compiled in LaTeX using the IFCoLog class file.

Contents

Editorial 1259
    Michael Gabbay and David Svoboda

ARTICLES

Modern Czech Logic: Voprěnka and Hájek, History and Background 1261
    Vítězslav Švejdar

**Finitist Consistency Proofs and the Impact of Gödel's Incompleteness
Theorems on Hilbert's Metamathematics** 1273
    Matthias Schirn

**Formalism and Structuralism, a Synthesis:
the Philosophical Ideas of H. B. Curry** 1301
    Jonathan P. Seldin

Resnik's Structuralism in Light of the History of Mathematics 1307
    Ladislav Kvasz

A Complex Problem For Formalists 1337
    Arezoo Islami

Formalism and Set Theoretic Truth 1365
    Michael Gabbay

A Non-structuralist Alternative to Formalism 1381
Danielle Macbeth

Are *Ante Rem* Structuralists Descriptivist or Revisionist Metaphysicians? How We Speak About Numbers 1399
Prokop Sousedík and David Svoboda

What Sort of Mathematical Structuralism is Category Theory 1417
Josef Menšík

Editorial Preface

Michael Gabbay
Department of Philosophy, University of Cambridge

Prokop Sousedík
Catholic Theological Faculty, Charles University

David Svoboda
Catholic Theological Faculty, Charles University

This special issue consists of papers deriving from some of the presentations at the conference "The Emergence of Structuralism and Formalism" held at Catholic Theological Faculty, Charles University, in June 2016. The conference brought together over 30 academics from Europe and North America. The conference presentations varied from historical discussions to new philosophical developments with respect both to Structuralism and Formalism.

This volume contains 9 papers that represent a cross-section of these presentations. We begin, appropriate to the location of the conference, with a survey of the history of Czech logic. We follow with two papers that open their discussions from the perspective of the historical development of Formalism and Structuralism, and then with two critiques of formalist and structuralist views in the light of the history of mathematics. We continue with two, more speculative, discussions of Formalism and finally two of Structuralism.

This volume reflects the diversity of the presentations at the original conference: some were technical, some historical and some more philosophical. What this volume cannot reflect, was the lively and scholarly interactions between the academics who attended, either as presenters or only as participants. However, we hope that such discussions will continue, and the subjects of Formalism and Structuralism remain as active fields of research as was exemplified in the conference.

We thank the Czech grant agency (GACR 18-05838S) for supporting us during the production of this volume.

Modern Czech Logic: Vopěnka and Hájek, History and Background

Vítězslav Švejdar*
Charles University, Prague
vitezslav.svejdar@cuni.cz

Abstract

We review some aspects of the history of modern Czech logic, with an emphasis on the role of Petr Vopěnka and Petr Hájek in the earlier stage of their careers.

Keywords: Petr Vopěnka, Petr Hájek, Set Theory, Nonstandard Numbers, Semiset.

1 Introduction

This paper is devoted to two distinguished Czech logicians, *Petr Hájek* and *Petr Vopěnka*. While Petr Hájek is well-known everywhere and has thousands of citations to his works in the WoS database, Petr Vopěnka is much less known. He did have some contacts in Poland and in Russia, but he did not travel much, and almost never left the Russia-dominated part of the world, and thus (as I was told by Petr Hájek) some of the people who knew his name even suspected that he might be a virtual person, much like Nicholas Bourbaki.

Both Hájek and Vopěnka had a major influence on modern Czech (and world) logic. In this paper we will review some of their results and achievements. We will also mention their motivations and some social aspects of their work, and try to say something about their philosophical background. Vopěnka, though he had fewer Ph.D. students and fewer citations than Hájek, was nonetheless a key figure in the history of logic. We think that his early work can serve as a subject matter of an interesting historical research project. This paper could be viewed as an early step in such research. However, what is written here is based more on personal memories than on serious study of sources.

---

*I would like to thank Bohuslav Balcar, Tomáš Jech, Peter Vojtáš, Derek von Barandy and Eva Ullrichová for useful comments, and for remarks that improve readability.

2 Some genealogy

Petr Hájek (1940–2016) was officially a student of *Ladislav Svante Rieger*, an algebraic logician who collaborated for example with Helena Rasiowa and Roman Sikorski. Among other things, Rieger invented the Rieger-Nishimura lattice (or Rieger-Nishimura ladder), a beautiful structure of infinitely many non-equivalent intuitionistic propositional formulas built up from a single atom: if the atom is p, then the bottom formulas in the ladder are $\bot$, p, $\neg p$, $p \vee \neg p$, $\neg\neg p$, $\neg\neg p \to p$, $\neg p \vee \neg\neg p$, ... The same structure was later independently invented by Nishimura. Rieger was also aware of the existence of nonstandard models of Peano arithmetic, which was by no means commonplace in the 1950s. Vopěnka was not sure about the details and origin of the proof known to Rieger, but thought that the construction was in fact Skolem's. A visitor of the Czech Republic may notice Rieger streets and Rieger park in Prague and in other cities; these are named after František Ladislav Rieger, who was active in the Czech national movement in the 19th century and was the great-grandfather of Rieger the mathematician.

Rieger (1916–1963) worked with both Vopěnka and Hájek, but he died soon after Hájek became his student. Hájek then collaborated with Vopěnka and never failed to mention that it was in fact Vopěnka who was his teacher.

Petr Vopěnka (1936–2015) would not call himself a logician: his original research interests, lasting in a sense through his entire life, were geometry and topology. He was a student (the last student) of Czech topologist *Eduard Čech*, who strongly influenced him and "showed him how to do mathematics". Čech is known by Čech cohomology, and some Czech mathematicians believe that it was he who invented Stone (or Čech-Stone) compactification. Vopěnka, when asked about it, once explained that as to Čech and Stone one cannot really tell which results should be attributed to whom: they worked together and were in friendly and frequent contact.

3 Vopěnka's set theory seminar

During the 1960s Vopěnka ran a seminar devoted to set theory. It was known as *Vopěnka's seminar* because Vopěnka was much more than an organizer. The seminar took place at the School of Mathematics and Physics of Charles University every Monday afternoon. Its participants usually waited to hear what Vopěnka was able to invent during the weekend and (I am told) during the late Sunday night, and he managed to stimulate research and suggest problems that they found meaningful.

The participants of the seminar were people who later became known as experts in set theory and in other fields, for example Bohuslav Balcar, Petr Hájek, Karel

Hrbáček, Tomáš Jech, Karel Příkrý, Antonín Sochor and Petr Štěpánek. Jiří Polívka and Oswald Demuth were not really set theorists, but it is important to mention them, too. Demuth studied in Russia and worked (very successfully) in constructivist mathematics done in the Russian style. He probably did not attend Vopěnka's seminar (indeed, set theory was not consistent with his philosophical views) but was in contact with Vopěnka and can be counted among his group. Polívka was a philosopher. In those times Vopěnka was not only a respected leader of a research group—at the School of Mathematics he was also one of the decision makers. He established a new department and became its head, founded a field of study in which the students of mathematics could specialize (its official title was "theoretical cybernetics", but to large extent it was logic), and he had a say in hiring. He wanted to have a philosopher (i.e., Polívka) in his group, and had the power to make it happen. Two members of the group, Hájek and Sochor, were officially affiliated with the Czechoslovak Academy of Sciences, which in those times was a research institution without study programs, but with the right to award Ph.D. degrees. The communist regime became more and more liberal in the 1960s. Nevertheless, Hájek and Sochor could not work at the university because they were known to be active Christians (an Evangelical and a Catholic respectively). Hájek even served as an organist in his church. The different official employers had no impact on the research interests of Hájek and Sochor, who normally participated in the work of the seminar.

Hájek was familiar with the logical literature and probably played the role of a set theorist who knew a lot about logic. In general, he put emphasis on contacts with logicians from Europe and the U.S., could fluently communicate in several languages and was sure that results should be presented at conferences or sent to journals published in English. All this was less true about Vopěnka, who in general preferred thinking to reading or to attending conferences. This is not meant to say that Vopěnka was worse in communication: for me personally, he was the best speaker among the professors I met at the Mathematical School in the 1970s. However, he was an excellent speaker and teacher only when speaking Czech.

A big stimulus for Czech set theorists was the continuum problem, which was unsolved in the early sixties. Gödel's proof of the consistency of the continuum hypothesis CH, employing the constructible universe, was known to them, but they were also aware that quite different model constructions would be needed to show the *unprovability* of CH. Some other tools were at hand: Vopěnka for example published an ultraproduct construction of a model of the Gödel-Bernays set theory **GB**. Thus when Paul Cohen solved the continuum problem in 1963, Prague set theorists were not unprepared. Vopěnka seldom attended conferences, but in 1963 he accidentally was present (in Nice, France, or maybe in the U.S., but some of his colleagues think that it was in Vienna) at a meeting where Cohen presented his result. While

Cohen probably considered his discovery a single-purpose proof and did not continue working on consistency results, people around Vopěnka were encouraged and started to think about turning Cohen's proof into a *method* that would yield further results. Such a method was established as soon as the following year, i.e. in 1964. It was called the ∇-*model* and it was a variant of the method of *Boolean valued models*. The same method was independently but somewhat later invented by Scott and Solovay.

Out of the axiomatic set theories, the Zermelo-Fraenkel set theory ZF is now more popular. However, Prague set theorists preferred to work with the Gödel-Bernays set theory GB, which is nowadays sometimes denoted NBG, where the letter 'N' refers to von Neumann. As the reader probably knows, the primary notion of ZF is *set*, and all other notions (number, function, ...) are reduced to it. Both theories are formulated in the same language $\{\in\}$ containing one unary predicate symbol for membership. Besides $\in$, the equality symbol $=$ can also appear in formulas (both theories are theories in predicate logic with equality). In GB, the primary notion is *class*, and sets are defined as those classes that are elements of other classes. It is common to use the uppercase letters $X, X_1, Y, \ldots$ to denote classes, and the lowercase letters x, y, etc. to denote sets. The axiom system of GB is usually presented as several axioms (e.g. the axiom of extensionality) and one schema, the *comprehension schema*. A *formula* of GB is *normal* if the class variables are not quantified in it. Here are four examples of normal formulas: $\neg \exists v(v \in x)$, $x \in Y_1 \mathbin{\&} x \in Y_2$, $x \notin x$, and $x = x$. The comprehension schema stipulates that every normal formula $\varphi(x, z_1, \ldots, z_k, Y_1, \ldots, Y_r)$, with a dedicated variable x and any number of set parameters $z_1, \ldots, z_k$ and class parameters $Y_1, \ldots, Y_r$, determines a class (of all sets x such that $\varphi(x, \underline{z}, \underline{Y})$); this class is unique by the axiom of extensionality. The classes determined by the four example formulas above are the empty class $\emptyset$, the intersection $Y_1 \cap Y_2$ of the classes Y_1 and Y_2, the class D of all sets that are not elements of themselves, and the universal class (of *all* sets, traditionally denoted V).

There is nothing paradoxical about the class D of all sets that are not elements of themselves; in GB, Russell's argument becomes a proof that D is not a set, i.e. that it is a *proper class*. The fact that a subclass of a set is again a set is provable in GB. Therefore, the universal class V is a proper class as well. Sets exist, and thus the empty class $\emptyset$, being a subclass of every class, is a set.

It is known that GB is conservative over ZF w.r.t. set sentences: any sentence not containing class variables is provable in GB if and only if it is provable in ZF. Therefore, ZF and GB can be seen as variants of the same theory that differ only inessentially. One can think that proper classes just simplify language in some cases. However, the relation between ZF and GB is quite involved, and actually the Prague group contributed to its clarification. One of their results will be outlined here.

A *truth relation* on a number n is a relation R between *set formulas* smaller than n and valuations of variables such that R satisfies Tarski's conditions (whenever applicable) with respect to the entire universe of sets. In more details, we can assume that variables that can appear in formulas are taken from a countably infinite set $\mathrm{Var} = \{v_0, v_1, v_2, \ldots\}$ and that we have coding of formulas using natural numbers. That is, formulas can be identified with their numerical codes. Among numbers smaller than n some are (codes of) formulas, i.e. some are Gödel numbers (of formulas). A valuation of variables is any function defined on the set Var. If R is a truth relation on n, then a pair $[v_i \in v_j, e]$, where $v_i \in v_j$ is a formula smaller than n and e is a valuation, is in R if and only if $e(v_i)$ is an element of $e(v_j)$. A pair $[\varphi \,\&\, \psi, e]$, where again $\varphi \,\&\, \psi < n$ and e is a valuation, is in R if and only if both pairs $[\varphi, e]$ and $[\psi, e]$ are in R, and similarly for the atomic formula $v_i = v_j$ and for other logical connectives and quantifiers. One can think of a truth relation on n as a table with finitely many lines (in the sense of GB, i.e. according to the definition of finite set formulated in GB) corresponding to those numbers smaller than n that are formulas, and with class-many columns corresponding to all valuations. The following facts can be proved in GB about truth relations: if a truth relation on n exists, then it is unique, a truth relation on 0 does exist, and if there exists a truth relation on n, then there also exists a truth relation on $n+1$. A number is *occupable*, denoted $\mathrm{Ocp}(n)$, if there exists a truth relation on n. Occupable number is a well defined notion in GB. However, the formula $\mathrm{Ocp}(n)$ is not normal (because it starts with the class quantifier $\exists R$), and thus in GB it is not guaranteed that it determines a class. It is however clear that if it determined a class, then the class would in fact be a set because it would be a subclass of the set of all natural numbers. The assumption that all natural numbers are occupable implies (provably in GB) the consistency of ZF. Since GB is conservative over ZF (provably in GB), it also implies the consistency of GB. To sum up, from Gödel's second incompleteness theorem we have $\mathsf{GB} \not\vdash \forall n \mathrm{Ocp}(n)$.

A formula $I(n)$ such that $\mathsf{GB} \vdash I(0)$ and $\mathsf{GB} \vdash \forall n(I(0) \to I(n+1))$ is called a *definable cut*. In GB we can have nontrivial definable cuts, i.e. definable cuts I such that $\mathsf{GB} \not\vdash \forall n I(n)$. A nontrivial definable cut violates induction. Therefore, full induction (for all formulas of GB, normal or not) is not provable in GB.

I believe that Vopěnka and Hájek were the first logicians who were aware of the existence of nontrivial definable cuts in GB, and that they discovered them well before 1973, when they published the above construction in [13].

Speaking roughly, one can think of a model (in the sense of the usual logical semantics, i.e. a set model) of GB as a model of ZF where some subsets are interpreted as classes. However, there could be subsets that cannot be accepted as classes without violating the axioms of GB. A model of ZF can contain a "subcollection" that is a subset from the metamathematical point of view, but it is neither a set nor a class in the sense of the model in question. And such a subcollection can even be included in some set of the model. This and similar observations led Vopěnka and Hájek to the notion of semiset and to a new version of axiomatic set theory, the Theory of Semisets TS, see [12]. A *semiset* is defined as a subclass of a set; a semiset is *proper* if it is not a set. TS is weaker than GB (in which no proper semisets exist), but it is still an extension of ZF. Therefore, it is a conservative extension of ZF w.r.t. set sentences. Proper semisets are just a possibility: their existence cannot be proved in TS, and GB can be obtained from TS by adding the axiom **every semiset is a set**. With semisets, some model constructions including those showing the independence of the continuum hypothesis could be more natural.

Vopěnka and Hájek believed that we (should) have the freedom to work with abstract axiomatic theories. Not only are these theories interesting because one can encounter exciting proofs when thinking inside them, but also they are useful because they provide a safe environment for all classical mathematics. However, Vopěnka and Hájek were finitists in the sense that they (in those times) also believed that on the metamathematical level, i.e. when reasoning *about* the axiomatic system, one has to be careful and use only those tools that are indisputable. In the beginning, the finitistic ideas could have been the reason to prefer working with GB: it is finitely axiomatizable. There is, however, a deeper reason for working with GB or TS, also of a finitist nature. When using the *method of forcing*, as it has developed and as it is now always presented, one works with a model of ZF and with a poset P in it, and with a generic filter on P. The filter is supposed (can by other means be shown) to exist, but cannot exist as an element of the given model. Reasoning about TS could be more economical as to the indisputable tools that are accepted on the metamathematical level: there is no distinction between a model and its outside, because the generic filter can be understood as a semiset existing in the model.

There are two results that may be worth mentioning when talking about set theories, Vopěnka, and Czech logic. In connection with category theory, Vopěnka invented (probably soon after 1964) a large cardinal axiom, now called *Vopěnka's principle*. It has several equivalent formulations, one of which being **every proper class of first order structures contains two different members such that one of them can be elementarily embedded to the other**. The principle is quite strong, stronger than other popular principles like, say, the existence of measurable cardinals (look at the cover of Kanamori's book [7]). Vopěnka did not really work in the field of large cardinals, and considered them a dead-end in mathematical research. For me personally the fact that he is the author of a widely known result in this area is evidence that he valued a good proof more than a methodological (philosophical, ideological) idea.

While ZF and GB prove the same set sentences, they are not equally efficient. Pavel Pudlák, who does not really belong to this section because he started to work with Petr Hájek later in the 1970s, used proof-theoretic methods to prove a speed-up theorem: eliminating classes from a GB-proof of a set sentence, i.e. constructing a ZF-proof from a given GB-proof of a set sentence, can cause a superexponential increase in the length of the proof.

4 Dissolution of (the classical) seminar

The Russian occupation of Czechoslovakia in August 1968 changed many things. In April 1969 the communist power started what they called 'normalization', i.e. restoration of the totalitarian regime, and these changes had an impact on Vopěnka's group. Vopěnka was identified as a non-cooperating person and lost his say in the administration of the School of Mathematics. He could still do research and had exciting ideas to follow, but his contacts with students were restricted and the department of which he was the head was disbanded. Some people (Jech, Příkrý and Hrbáček) emigrated to the U.S., but some other (younger) people showed interest in working with Vopěnka. The communists did not go so far as to fire Vopěnka from the School of Mathematics, but the threat was there (a few years later they fired Polívka, the philosopher mentioned above).

In this situation Vopěnka told his colleagues that they had learnt everything they could from him, and thus they should not rely on him and find their own topics to work on; he himself would start a new field of research with a new group of collaborators. Vopěnka evidently aimed to stimulate the intellectual development of his colleagues. However, there was also patriotic reasoning behind his decisions: in the times when contacts with the world were violently broken, doing something specifically Czech was his way of maintaining the nation's culture.

For Balcar and Štěpánek the "new" topic consisted in continuing their work in classical set theory. They also published a very influential (Czech) textbook in set theory. Balcar also worked with the topologists Petr Simon and Zdeněk Frolík, while Štěpánek was active in the study programs of the newly created Department of Theoretical Computer Science. Petr Hájek started to work on two different things mentioned below in more detail, the GUHA method on one side, and metamathematics of arithmetic on the other side. Sochor was the only member of the classical seminar whom Vopěnka accepted to his new group (or, as some insiders describe it, who was not sufficiently obedient to follow Vopěnka's advice to find another topic).

Vopěnka's new (the specifically Czech) research field was the Alternative Set Theory, and his new group (new seminar) included for example Karel Čuda, Josef

Mlček, Alena Vencovská, Kateřina Trlifajová (and Sochor). Some axioms and definitions of the *Alternative Set Theory* AST are taken from GB: classes, sets, extensionality, ordinals. In some aspects AST is similar to the theory of finite sets: no limit ordinals exist, and the class N of all natural numbers is a proper class (indeed, it equals the class of all ordinals). The class N contains a proper initial segment FN of all finite natural numbers. The class FN is a proper semiset, numbers in N − FN are infinite (nonstandard) natural numbers. There is some freedom in choosing axioms concerning bijections and cardinals. In the most popular and the most natural version, however, for every class X there is a bijection that maps X either on some finite natural number, or on the class FN, or on the entire universe V (of all sets). In the latter case X has cardinality continuum. In this sense there are only two infinite cardinals in AST, the countable infinity and the continuum. Large parts of mathematics can be recovered in AST. However, some parts, like category theory or functional analysis, are problematic or impossible in AST. What can be recovered is *topology*, and what looks very elegant and simple in AST are notions like real numbers or continuous functions, and the differential and integral calculus. This is because it holds that if α is an infinite natural number (a number from N − FN), then $\frac{1}{\alpha}$ and $-\frac{1}{\alpha}$ are infinitely small numbers, i.e. *infinitesimals*. Then, for example, a real function f is continuous if $|f(x + \delta) - f(x)|$ is infinitesimal for every infinitesimal δ. Thus AST can be seen as a suitable theory for *nonstandard analysis*. Vopěnka liked to emphasize that quite a substantial amount of mathematics existed before set theory, and that "mathematical analysis" was obtained by shortening the term "mathematical analysis of infinitesimals", used in the times of Leibniz.

The GUHA method, one of the activities of Petr Hájek after the dissolution of the classical seminar, combined statistical and logical tools to automatically discover dependencies in data, with an emphasis on anomalous dependencies shared by only a small fraction of the data. The data were typically medical data (concerning health of patients), but GUHA was of course a general method applicable to data of various kinds. It would now be described as a data-mining method. Hájek worked on it with a group of researchers of different affiliations; they met regularly at a *seminar on applied logic*. The group included Tomáš Havránek, and the book [3] is far from being the only publication about this method.

Another big field of Hájek's research interests lied in *metamathematics of arithmetic*, for which an important initial source of information was Feferman's paper [1]. Besides other things, Hájek studied the interpretability of axiomatic theories. He noticed that the two closely related set theoretic systems ZF and GB differ in interpretability. To look at this in more detail, let $S \triangleright T$ be a shorthand for "T is interpretable in S", and let $\mathrm{Intp}(T)$ be the set $\{\,\varphi\,;\,T \triangleright (T + \varphi)\,\}$ of all sentences of T whose consistency with T can be shown using an interpretation. It is easy to see that Intp(GB) is a recursively enumerable (r.e.) set. Hájková and Hájek proved in [5] that if A is an r.e. set of set sentences such that no member of A is refutable in ZF (note that this condition is satisfied by the set Intp(GB), and note also that a sentence is not refutable in GB if and only if it is not refutable in ZF), then the set Intp(ZF) $- A$ is nonempty. It immediately follows that Intp(ZF) $\neq$ Intp(GB) and that Intp(ZF) is not r.e. Yet another result of Hájek, somewhat less involved but important because it later (in [9]) translated to one of the axioms of interpretability logic, should not be forgotten: if $(S+\varphi) \triangleright T$ and $(S+\psi) \triangleright T$, then $(S+\varphi\vee\psi) \triangleright T$.

Hájek assembled a small group working on this field, which for a while (in 1977) included his wife Marie, Kamila Bendová and myself. The group was soon moved to the Institute of Mathematics (of the Czechoslovak Academy of Sciences), turned into the *seminar on metamathematics of arithmetic*, or simply *Logic Seminar*, and was joined by Pavel Pudlák, whose contribution was enormous. Hájek inspired research in fragments of arithmetic, interpretability of theories and in interpretability logic. This research definitely had an international dimension, which is not quite true about about GUHA or about Vopěnka's AST. For example, the unpublished but often cited Solovay's paper [8] (which used the idea of occupable numbers) answered positively a question asked in [5], i.e. whether Intp(GB) $-$ Intp(ZF) is nonempty. For Petr Hájek, this period ended in the early 1990s when the book [4] appeared. He then moved to another research field: he started an extensive research in *fuzzy logics* and established an active group of researchers working on this topic. The Logic Seminar, with Pudlák and with other distinguished logicians, but without Hájek, and with topics like proof complexity, continues to be very active.

5 Some more motivations and background

In a certain sense, and at least in the earlier period, Vopěnka and Hájek were formalists: they supposed that we work in axiomatic theories and that truth in these theories is determined by their axioms. The first of the following two quotes from the Introduction to [12] (written, just as most of the book, by Hájek) contains "game with symbols", a term that is sometimes taken as evidence of *formalism*.

> "To prove a statement about sets, classes, etc. we appeal to the axioms of the theory in question. On the other hand, statements about statements or proofs are metamathematical statements. [...] In the metamathematical investigation, a theory is to be treated as a purely formal game with symbols."

> "We shall exclude certain methods of proof, such as proof by contradiction, from metamathematical arguments. If we assert that a certain

proof exists, then we shall always give instructions for constructing it. [...] Our methods can be said to be finitistic."

However, the expression "game with symbols" cannot be understood literally. We need axioms to have something reliable, but the choice of axiomatic theories is not arbitrary because these theories are expected to reflect the classical mathematics and serve as an environment for it. When choosing the axioms, we know or at least have a feeling what proofs and concepts should be formalizable in the resulting systems, and the systems should not be unnecessarily strong. The word "finitistic" in the second quotation may be understood as "constructivist", and the whole quotation calls for using intuitionistic logic on the metamathematical level. However, when I asked Hájek around 1977 whether he really thought that intuitionistic logic was the adequate logic for reasoning about axiomatic theories, his answer was no. His position was then roughly as follows. Whenever we have a mathematical proof, we can ask the question what is the formal system in which the proof can be formalized. The system can be weak or not, but "weak" is not the same as "intuitionistic"; we in fact have no good reason to prefer intuitionistic logic to classical logic.

Vopěnka considered finite natural numbers the "visible" part of the universe. The idea that whatever can be observed when looking at the visible world must have at least some continuation behind the horizon is inspired by *phenomenology*. One of Vopěnka's sources of philosophical (phenomenological) thinking was Jiří Polívka. He was a student of Czech philosopher Jan Patočka, who during his studies worked with Heidegger, Husserl and Eugene Fink. A somewhat unrelated remark about Polívka: When he was fired from the School of Mathematics, he accepted a job working at a warehouse, which was not very time consuming, and he could do what he thought was his task—edit the nachlass of Jan Patočka. For Vopěnka, another source of inspiration was *theology*. It sounded like a joke when he lectured about the Greek gods who are not so powerful as the Christian God, but he was serious: he believed that the idea of the variety of infinite cardinal numbers, present in the classical set theory, emerges from theological thinking.

While Petr Hájek was active in the Evangelical Church of Czech Brethren, he separated this activity from his mathematical work. In research, he did not seem to be inspired by religion or by philosophy, and was rather pragmatic when choosing the problems to think about: a good research field for him was one in which it is possible to prove strong and interesting theorems.

References

[1] S. Feferman. Arithmetization of metamathematics in a general setting. *Fundamenta Mathematicae*, 49:35–92, 1960.

[2] P. Hájek. *Metamathematics of Fuzzy Logic*. Kluwer, 1998.

[3] P. Hájek and T. Havránek. *Mechanizing Hypothesis Formation (mathematical foundations for a general theory)*. Springer, 1978.

[4] P. Hájek and P. Pudlák. *Metamathematics of First Order Arithmetic*. Springer, 1993.

[5] M. Hájková and P. Hájek. On interpretability in theories containing arithmetic. *Fundamenta Mathematicae*, 76:131–137, 1972.

[6] Z. Haniková. Petr Hájek: A scientific biography. In F. Montagna, editor, *Petr Hájek on Mathematical Fuzzy Logic*, number 6 in Outstanding Contributions to Logic, pages 21–38. Springer, 2015.

[7] A. Kanamori. *The Higher Infinite*. Springer, 2nd edition, 2003.

[8] R. M. Solovay. Interpretability in set theories. Unpublished letter to P. Hájek, Aug. 17, 1976, http://www.cs.cas.cz/~hajek/RSolovayZFGB.pdf, 1976.

[9] V. Švejdar. Modal analysis of generalized Rosser sentences. *J. Symb. Logic*, 48(4):986–999, 1983.

[10] P. Vopěnka. *Mathematics in the Alternative Set Theory*. Teubner, Leipzig, 1979.

[11] P. Vopěnka. Prague set theory seminar. In P. Cintula, Z. Haniková, and V. Švejdar, editors, *Witnessed Years: Essays in Honour of Petr Hájek*, Tributes, pages 5–8. College Publications, London, 2009.

[12] P. Vopěnka and P. Hájek. *The Theory of Semisets*. North-Holland and Academia, Praha, 1972.

[13] P. Vopěnka and P. Hájek. Existence of a generalized semantic model of Gödel-Bernays set theory. *Bull. Acad. Polon. Sci., Sér. Sci. Math. Astronom. Phys.*, XXI(12), 1973.

Finitist Consistency Proofs and the Impact of Gödel's Incompleteness Theorems on Hilbert's Metamathematics

Matthias Schirn
Munich Center for Mathematical Philosophy, University of Munich

Abstract

In this article, I first discuss several issues in Hilbert and Bernays's extension of the finitist point of view in the second volume of *Foundations of Mathematics* (1939). In what follows, I argue that by appreciating the distinctive character of the notion of an "approximative" consistency proof we can make good sense of the nature of the consistency proofs that Hilbert outlines both in his classical papers on proof theory in the 1920s and in the first volume of *Foundations of Mathematics* (1934). I further argue that a weak version of Hilbert's programme is compatible with Gödel's second incompleteness theorem by using only what are clearly natural provability predicates.

Keywords: Hilbert's Programme, Finitism, Formalism, Consistency Proof, Incompleteness Theorem, Proof Theory

1 Introduction

In the 1920s, Hilbert developed his finitist proof theory in order to defend classical mathematics by means of an unassailable metamathematical consistency proof. The key idea underlying such a proof was to establish the consistency of a mathematical theory T by means of weaker, but at the same time more reliable methods than those that could be formalized in T. It was in the light of Gödel's incompleteness theorems that finitist metamathematics as designed by Hilbert in the 1920s turned out to be too weak to lay the logical foundations for a significant part of classical mathematics. Contrary to what some scholars working on Hilbert's programme assume, Hilbert's metamathematics of the 1920s was possibly weaker than Primitive Recursive Arithmetic (PRA). In any event, there is, as far as I can see, no textual evidence in his classical papers on proof theory that he identifies finitist metamathematics with

PRA.[1] In the 1930s, Hilbert and Bernays responded to Gödel's challenge by extending twice the original finitist point of view.[2] In the first volume of *Grundlagen der Mathematik (Foundations of Mathematics)* [29], metamathematics is taken to be at least as strong as PRA and may include even PRA + Con$_{\text{PRA}}$, while in the second volume the authors seem to take metamathematics to be as strong as Peano arithmetic (PA) augmented by transfinite induction up to ε_0. The extensions (especially the second) were guided by two central, though possibly conflicting ideas: first, to make sure that they preserved the quintessence of finitist metamathematics; second, to carry out, within the extended proof-theoretic bounds, a finitist consistency proof for a large part of mathematics, preferably for second-order arithmetic.[3]

In his classical essays on proof theory in the 1920s, Hilbert characterizes metamathematics as a set of statements about figures or strings of figures drawn from a finite stock. Unlike the mathematical objects which, from a non-formalist point of view, are usually conceived of as abstract (that is, as non-spatial, non-temporal, not involved in causal interaction) the figures considered in metamathematical operations are concrete, spatio-temporal objects and as such intuitable and — in the

---

[1] The design of PRA seems to be motivated by specifically metamathematical reasoning. PRA is framed in a quantifier-free fragment L_{PRA} of a first-order language. The vocabulary of L_{PRA} contains for every primitive recursive function f a unique function-sign "F". The axioms of PRA are, apart from classical logic in L_{PRA}, the recursion equations for all primitive recursive functions. PRA is not finitely axiomatizable and it is closed under the induction rule.

[2] W.W. Tait [65] criticizes me and my co-author K.-G. Niebergall for our claim (in [46]) that Hilbert's conception of finitism in [29] extends the original conception that Hilbert propounded in the 1920s. In [55], I reject Tait's criticism as unjustified. A general, although not clear-cut, characterization of the finitist point of view is given in [29, p. 32]. Regarding Hilbert's finitism see [72], [65] and [63]. In [56, 57], it is argued that Tait's thesis (in [64]) "The finitist functions are precisely the primitive recursive functions" is disputable and that another, likewise defended by him in [64], is untenable. The second thesis is that the finitist theorems are precisely the universal closures of the equations that can be proved in PRA. See the discussion of Tait's conception of finitism from another perspective in [32, §3-§4]. Incurvati argues that there are at least two versions of finitism which "seem to pull in different directions" (p. 2434); see in this connection also [11]. For a defence of strict finitism see [68].

[3] Since I have only a fairly limited space in this paper, I cannot adequately discuss some interesting ideas in [8] (see the useful comments in [42]) nor can I appropriately comment on [39, 7, 10] and [58, 59, 60]. I hope to be able to do this in the very near future. I think that [60] provides the thoroughest and most comprehensive investigation of Hilbert's programme that has been published until today. For a much shorter, but likewise clear, scholarly and instructive account of Hilbert's programme "then and now", see [73]. Those who are interested in the development of contemporary proof theory and its applications in mathematics starting with Hilbert's programme should carefully study [38]. For an interesting account of the relation between the fundamental ideas of Hilbert's programme and critical philosophy in the splendid setting of the "Göttinger Schule" around Hilbert see [50]; see in this respect also [51] and [41, pp. 165 ff.] where mainly Oscar Becker's arguments against Hilbert's view on mathematical induction are discussed.

ideal case — surveyable. The axioms, formulae and proofs of formalized arithmetic are on a par with the numerical figures that in the early 1920s Hilbert constructs by means of intuitive methods. They are the subject-matter of the contentual, proof-theoretic considerations. For Hilbert, a formalized proof, like a numerical figure, is a concrete and surveyable object which must be given as such to our perceptual intuition. The term "(natural) number" is reinterpreted. It is explained by means of the term "sign", at least in [22]. If we allow a massive dose of anachronism, the procedure suggested by Hilbert could thus be seen as reversing the post-Gödelian method of coding signs by numbers.

Furthermore, Hilbert's metamathematics of the 1920s is neither formalized nor axiomatized. It is supposed to allow only intuitive and contentual reasoning, and, as he and Bernays stress (cf. [29], p. 32), this kind of reasoning is, by its very nature, free of axiomatic assumptions. Due to the sharp distinction between formalized mathematics and informal metamathematics, the meanings attached to the word "to prove" in the two disciplines are strikingly different. In formalized mathematics, it means *to infer according to the formal rules of the calculus*; in contentual meta-mathematics, it means *to show by means of contentual, intuitively evident inference*. Again, it is precisely the intuition-based character of metamathematical reasoning that is supposed to guarantee its security and reliability.[4]

Contentual, informal metamathematics serves the purpose of carrying out a finitist consistency proof for a given formalized mathematical theory T. The consistency claim for T is said to amount to the claim that there is no proof of "$1 \neq 1$" in T which, according to Hilbert, can be established within the scope of intuition, if T is in fact consistent.[5]

---

[4] In [48], Parsons analyzes the relation between finitist proof and intuitive knowledge (see also the chapter 'Intuitive arithmetic and its limits' in [49] which overlaps considerably with [48]). The thesis that a proof of a sentence S according to the finitist method provides intuitive knowledge of S he calls Hilbert's Thesis. In Parsons's view, the elementary axioms of PRA are intuitively known if they are interpreted relative to Hilbert's intuitive model of strings. In particular, Parsons argues that the logical inferences employed in PRA preserve intuitive knowledge and that induction can reasonably be understood to do this. Primitive recursion is said to render a defence of Hilbert's Thesis rather difficult. However, Parsons maintains that addition and multiplication can be seen intuitively to be well defined. After having discussed some views in [29], Parsons concludes that the authors have failed to advance a non-question begging argument for the claim that all primitive recursive functions can be considered intuitively to be well defined. He adds though that there is no clear case against Hilbert's Thesis either. See the critical discussion of Parsons's position in [32, §2].

[5] The fundamental idea of Ackermann's consistency proof for a fragment of arithmetic in his 1925 article 'Begründung des "tertium non datur" mittels der Hilbertschen Beweistheorie' [1] is this: To show that within the proof figure the "ε" can be replaced by number signs in such a way that all formulae of the proof figure are turned into formulae that can be shown to be correct (richtig) according to the methods of finitist mathematics. What Ackermann refers to as "Gesamtersetzung"

2 Metamathematics in Hilbert and Bernays' *Foundations of Mathematics*, vol. 2, [29]

In his Introduction to the second volume of *Foundations of Mathematics* [30], Bernays points out that two main issues are on the agenda. The first is to give an account of Hilbert's proof-theoretic investigations which rest on methods linked with the ε-symbol including the method of eliminating the bound variables via the ε-symbol as well as the application of the ε-symbol to the investigation of the logical formalism. The second issue is the need to extend the finitist point of view in the light of Gödel's incompleteness theorems. The requisite extension leads Hilbert and Bernays to a consideration of Gentzen's purportedly finitist or at least constructive consistency proof for Peano Arithmetic (PA) in his seminal paper 'Die Widerspruchsfreiheit der reinen Zahlentheorie' ('The Consistency of Elementary Number Theory', [12]).[6]

Here I must confine myself to making a few comments on the ε-symbol in [30], §§1-3. In my view, the long-winded discussion is not always as clear as it should be.[7] In any event, the ε-symbol in its role in [30] is a kind of generalization of the μ-symbol for an arbitrary domain of individuals. The μ-symbol is a function sign which has a formula variable as argument that itself has an argument. As to the ε-symbol, it is, roughly speaking, a function of a variable predicate which apart from the argument that is linked to the bound variable and belongs to the ε-symbol, can also contain free variables as arguments ("parameters"). The value of this function

---

plays a central role in his consistency proof; see his definition of the term "Gesamtersetzung" in §6, pp. 20 f. In the last part of his paper 'Zur Hilbertschen Beweistheorie' (1927), von Neumann makes some comments on Ackermann's consistency proof ([69], pp. 44 ff.). He observes that Ackermann's formalistic approach has initially the same extension as his own, that is, "that with its help the same parts of mathematics can be grounded as with our own" (p. 44). He goes on to point out that Ackermann's consistency proof is not a gapless proof; see his arguments on pp. 44 ff. In his 1935 overview article ([3], pp. 213 f.), Bernays lists several finitist consistency proofs that have been carried out for subsystems of number theory by Ackermann in [1] (1925), von Neumann in [69] (1927), Herbrand in [20] (1931) and Gentzen in [12] (1936). He designates with "F_1" the system that one obtains from the first-order predicate calculus by adjoining the axioms of equality and of number, but by restricting the application of mathematical induction to formulae without bound variables. F_2 is the formalism that results from F_1 by adding to F_1 the ε-symbol and the ε-axiom. The consistency of F_2 was proved by Ackermann in [1] (1925) and von Neumann in [69] (1927); the consistency of F_1 was proved by Herbrand in [20] (1931) and Gentzen in [12] (1936). It seems that in his notes on Ackermann and von Neumann, Hilbert possibly overestimated both the scope and the significance of their consistency proofs.

[6] So far relatively little work has been published on [29]; see the discussions in [60], sections II.1 and II.5 and in [55].

[7] There can hardly be any doubt that the entire book was written by Bernays in his occasionally heavy-going German prose. But he most likely consulted Hilbert in the course of writing the book.

for a determinate predicate A is an object of the domain of objects. According to the contentual interpretation of the formula

$$(\exists x)A(x) \to A(\varepsilon_x A(x)) \tag{1}$$

this object is one to which the predicate A applies, provided that it applies to any object in the domain at all. To avoid possible objections to their treatment of the ε-symbol, Hilbert and Bernays regard the operations with it only as an *auxiliary* calculus from which significant benefit for metamathematical considerations can be derived. However, from the point of view of metamathematical investigations, they suggest eliminating the existential quantifier in (1). Thus, the resulting simpler formula

$$A(a) \to A(\varepsilon_x A(x)), \tag{2}$$

which they call the ε-formula, is laid down as the axiom that governs the ε-symbol. Thanks to the introduction of this axiom, the use of the ε-symbol in a given formalism is said to achieve three things:

1. Together with a couple of definitions (cf. [30, p. 15]), it provides the formulae and schemata for the quantifiers;

2. it replaces the ι-symbol (regarding the introduction and the use of the ι-symbol see [29, §8]), and

3. reduces the "symbolic dissolutions" to explicit definitions (cf. [30, p. 17]).

I shall now turn to the extension of the scope of finitism in the light of Gödel's incompleteness theorems. Perhaps somewhat surprisingly, Gödel himself was rather cautious as far as the impact of his second incompleteness theorem on Hilbert's programme was concerned. In [16], Gödel points out that his second incompleteness theorem does not contravene Hilbert's formalistic point of view. According to Gödel, it is conceivable that there are finitist proofs that cannot be expressed in a formal system P, where P is essentially the system obtained when the logic of Russell and Whitehead's *Principia Mathematica* is superposed upon the Peano axioms. Unfortunately, in his paper Gödel failed to provide any clue as to which finitist methods, that resist formulation in P, he had in mind.[8]

---

[8] In his letter to Gödel of 20 November 1930, von Neumann mentions his discovery of the unprovability of the consistency of mathematics, thus seemingly anticipating Gödel's second incompleteness theorem (cf. [70], pp. 123 f.) He observes that in a formal system S that contains arithmetic one can express that "1 = 2" cannot be the end formula of a proof which proceeds from the axioms of S. In fact, this statement is taken to be a formula of S; von Neumann calls it W and claims to

As regards the intended extension of the proof-theoretic means, it is worth noting that even before the publication of Gödel's 1931 paper Hilbert had introduced an inference rule whose status as a finitistically admissible device was debatable, namely a restricted ω-rule, henceforth referred to as to as "ω^*". The introduction of this rule was primarily intended as a means of extending PA, though it was perhaps also seen as a candidate for strengthening the inferential repertoire of metamathematics.[9] Rule ω^* reads as follows:

When it is shown that the formula $A(z)$ is a correct numerical formula for each particular numeral z, then the formula $\forall x A(x)$ can be taken as a premise.

Hilbert characterizes ω^* expressly as finitist and emphasizes that $\forall x A(x)$ has a much wider scope than $A(n)$ where n is an arbitrarily given numeral. However, he does not use ω^* in a metamathematical consistency proof. Now recall that finitist metamathematics of the 1920s does not include any inference rules in the ordinary sense. Finitist derivations are carried out either in the form of thought-experiments on intuitively given objects (cf. [29], p. 32) or simply with concrete objects. Thus, even if Hilbert had succeeded in justifying ω^* as a finitist inference rule, adjoining ω^* to the informal, purely contentual modes of inference available in metamathematics would have been at odds with the methodological underpinnings of his finitist metamathematics in the 1920s. Moreover, there is no textual evidence for the assumption that by 1930 Hilbert may have significantly changed his mind regarding the nature of metamathematics. It seems that at this time he still endorsed by and large the position that he had developed in his earlier papers on proof theory.

Let us return to [30]. If we compare, from the point of view of finitism, the ideas developed in this book with Hilbert's essays on proof theory in the 1920s, we notice a remarkable shift of emphasis. Genuinely philosophical reflections on the finitist nature of metamathematics recede by and large into the background, while the idea that consistency proofs for more extensive fragments of mathematics are

---

have shown that W is always unprovable in consistent systems, that is, "a putative effective proof of W could certainly be transformed into a contradiction" (p. 123). For Gödel, this was nothing new. In November 1930, he had already found his second incompleteness theorem and in his reply to von Neumann's letter of 20 November probably had informed him about this. (Unfortunately, Gödel's letter seems to have been lost.) From this theorem, von Neumann concludes — in his letter to Gödel of 29 November 1930 (cf. [70], pp. 124 f.) — that there is no rigorous justification of mathematics. However, he does not explain, let alone justify, this conclusion. He might have reasoned as follows: If we succeeded in carrying out a consistency proof for all of mathematics in metamathematics, mathematics would prove its own consistency, provided that metamathematics is only a (small) fragment of mathematics in its entirety. Yet the possibility that mathematics proves its own consistency is ruled out by Gödel's second incompleteness theorem, assuming that mathematics is recursively enumerable.

[9]See [26, p. 491].

actually to be carried out comes to the fore. It seems to be of lesser importance to ponder thoroughly over their finitist admissibility. The word "finitist" is still there, and is indeed used frequently by Hilbert and Bernays, but it seems to have lost a great deal of its original impact on the character of metamathematical reasoning. Nevertheless, in fairness to Hilbert and Bernays I should also mention that there are some scattered remarks in the book which seem to conform to the characterization of the finitist point of view, say, in 'On the Infinite'. We are told, for example, that the truth-value of a logical function for a given set of arguments is, in general, nothing that is finitistically determined, since in the definition of the logical function totalities and existential specifications may go in [30, pp. 197 f.]. As to the finitist interpretation of the quantifiers, we are referred to the account given earlier in [29, pp. 32 ff.]:

1. A general statement about numerals can be interpreted finitistically only as a hypothetical statement, that is as a statement about every concretely given numeral.[10]

2. An existential statement about numerals must be construed as a partial proposition, that is, as a statement which consists either in the direct specification of a numeral with the property $A(n)$ or in the specification of the procedure of obtaining such a numeral.

3. The combination of a general statement with an existential statement, say, 'For every numeral k with the property $A(k)$ there is a numeral l for which $B(k,l)$ holds" is explained in quite similar terms. Such a statement must be construed as the incomplete communication of a procedure by means of which we can find for each given numeral k with the property $A(k)$ a numeral l which stands to k in the relation $B(k,l)$.

Much later in [30] (cf. p. 358) the authors claim that a finitist assumption relates always to an intuitively characterized situation and that the truth of a general sentence cannot be considered to be such a situation. So much for the remnants of Hilbert's original finitism that we find in [30].[11]

---

[10] Hilbert and Bernays's interpretation of universal quantification is reminiscent of Gentzen's and Tait's accounts [12, 64]. The latter likewise embody a version of the ω-rule which rests on the identification of numerals with numbers. Tait's additional idea is that the law in question is to be regarded as something that is given by a finitist function.

[11] When Hilbert and Bernays raise the question of whether it is possible, with finitist means, to go beyond the scope of the modes of inference formalizable in Z_μ they admit that this question is not precisely formulated ([30], p. 361). (See the introduction of the number-theoretic formalism Z_μ in [30] in §5, 2 entitled 'The formalized metamathematics of the number-theoretic formalism', pp.

In this influential book, metamathematics is stated in a formal mathematical language and it is treated as an axiomatized theory which in the light of the work of Gödel, Gentzen and others does not come as a surprise. There is a section entitled "The arithmetization of metamathematics" in which the authors deal, for example, with the arithmetization of the concept of a formula, the concept of a derivation and also with the arithmetization of the distribution of truth-values (cf. [30], §4, pp. 215 ff.). Metamathematics is now at least as strong as PA (cf. [30], pp. 353 ff., 361 ff.), but following Gentzen, Hilbert and Bernays sympathize with the idea of making the proof-theoretic repertoire even much stronger than the inferential resources available in PA.

Like Gentzen [12], Hilbert and Bernays regard metamathematics as a formalized mathematical theory whose consistency must be presupposed. Roughly speaking, Gentzen's consistency proof for PA rests on a reduction procedure for sequents and derivations.[12] In order to prove the requisite finiteness of the reduction procedure, Gentzen attempts to show that each reduction step in a certain sense simplifies a given derivation. To this end, he assigns to each derivation an ordinal number that represents a measure for the complexity of the derivation. It is then shown that in a reduction step the ordinal number of a derivation usually diminishes. The key idea of this reduction method is that it enables us to recognize a simplification of the derivation in a reduction step despite the apparent increase in complexity. Finally, Gentzen proves the theorem of transfinite induction which he formulates in terms of the allegedly constructive notion of the accessibility of all ordinal numbers. To the complete system of ordinal numbers used by him corresponds Cantor's first ε-number of the second number class.[13] The first number class is the set of all finite ordinal numbers $\{\nu\}$, which has the type ω (the smallest transfinite ordinal number) while the second number class $Z(\aleph_0)$ is the set $\{\alpha\}$ of all order types α of well-ordered sets of cardinality $\aleph_0$ and thus the set of all transfinite denumerable ordinal numbers (cf. [5], pp. 325, 331).[14]

---

302 ff.). The lack of precision is said to be due to the fact that "finitist" has not been introduced as a sharply defined term, but only as a label for a methodic guideline which enables us to recognize certain kinds of concept formation and of inference definitely as finitist and certain others definitely as non-finitist. However, according to Hilbert and Bernays this guideline does not yield an exact dividing line between kinds of concept-formation and of inference which meet the requirements of the finitist method and those which do not.

[12] Concerning Gentzen's original consistency proof see [4], [34] and [67]. Regarding his second consistency proof see [52].

[13] Cf. Cantor's 'Beiträge zur Begründung der transfiniten Mengenlehre' ('Contributions to the Foundation of Transfinite Set Theory'), in [5], pp. 347 ff.

[14] The smallest of all ε-numbers is $\varepsilon_0 = E(1) = \omega_\nu$, where $\omega_1 = \omega$, $\omega_2 = \omega^\omega$, $\omega_3 = \omega^{\omega^2}$, ..., $\omega_\nu = \omega^{\omega_{\nu-1}}$ (see the proof in [5, p. 348]). The set of all ε-numbers of the second number class

Gentzen's theorem reads as follows ([12, p. 555], [15, p. 192]):

> All ordinal numbers are accessible in the following sense, by our running through them in the order of their increasing magnitude; the first number 0 is considered *accessible*; if further all numbers smaller than a number β have been recognized as *accessible*, then β is also considered *accessible*.

By virtue of this theorem, the finiteness of the reduction procedure for arbitrary derivations is said to follow at once (cf. [12, p. 556], [15, p. 192]). The finiteness of the reduction procedure is supposed to carry over from the totality of the derivations with the ordinal numbers smaller than β to the derivations with the ordinal number β. According to Gentzen's theorem, this property therefore holds for all derivations with arbitrary ordinal numbers. The definition of the concept *accessible* is specified as entirely constructive, on the ground that β is *accessible* only when all numbers smaller than β have previously been recognized as effectively accessible numbers. Gentzen emphasizes though that the word "all" used in this context must be interpreted finitistically since "in each case, we are dealing with a totality for which a *constructive* rule for generating all elements is given" ([12, p. 558], [15, p. 195]).

So, Gentzen argues in favour of the finitist admissibility of transfinite induction up to ε_0 — TI$[\varepsilon_0]$— by appeal to its purportedly constructive character. I think, however, that from a finitist point of view his argument rests on an overly liberal interpretation of universal quantification and that it does not carry conviction precisely for this reason. It seems that at least for formulae of restricted complexity Gentzen assumes that we can conclude to $\forall x A(x)$, if for every given numeral n we have already proved $\psi(n)$. Yet the purportedly finitist consistency proof for PA could at best work at the expense of accepting a version of the ω-rule which in turn would have to be justified as a finitist mode of inference. Since I fail to see how this could be done persuasively, I conclude that Gentzen has not succeeded in justifying TI$[\varepsilon_0]$ as a finitistically indisputable method of proof.[15]

---

forms, in its order of magnitude, a well-ordered set of type Ω of the second number class, conceived of in its order of magnitude, and has therefore the power $\aleph_1$ (cf. [5, p. 349]; see the proof of this theorem on pp. 349 f.).

The system of ordinal numbers used by Gentzen is well-ordered through the relation $<$. To the numbers with the "numerus" 0, 1, 2, 3, ... etc. correspond the transfinite ordinal numbers $\omega + 1$, $2^{\omega+1} = \omega + \omega$, $2^{\omega+\omega} = \omega \cdot \omega$, $2^{\omega \cdot \omega} = \omega^\omega$, etc. As was said above, to the complete system corresponds Cantor's first ε-number of the second number class.

[15] While Gentzen [12] attempts to establish the finiteness of a certain reduction procedure, Ackermann [2] intends to establish the finiteness of the succession of what, by picking up a term already employed in [1], he calls "Gesamtersetzungen" [2, pp. 175 ff.]. Like Gentzen, Ackermann makes use of TI$[\varepsilon_0]$. However, unlike Gentzen, Ackermann does not pretend that his consistency proof for full

On reflection, it is clear that from Gödel's second incompleteness theorem and the fact that Gentzen [12] and [13] proved the consistency of pure number theory (PA) by employing only purely number-theoretic means plus $TI[\varepsilon_0]$ one can indirectly draw the following conclusion: $TI[\varepsilon_0]$ cannot be proved with purely number-theoretic proof means. For suppose that $TI[\varepsilon_0]$ were provable in PA. A consistency proof for PA could then be carried out with the proof-theoretic means available in PA. But this would clash with Gödel's second incompleteness theorem.

In his *Habilitationsschrift* [14], Gentzen proves the unprovability of $TI[\varepsilon_0]$ in PA in a direct way. Thanks to the method applied in this proof, he is able to prove in addition that certain "initial cases" (*Anfangsfälle*) of transfinite induction up to numbers below ε_0 are unprovable in certain fragments of PA. This procedure bears a strong resemblance to his consistency proof for PA, since what is actually carried out in the two cases is a proof of unprovability. Furthermore, the claim that $TI[\varepsilon_0]$ is unprovable in PA presupposes the consistency of PA. For if PA were inconsistent, every sentence could be proved in PA, and thus also $TI[\varepsilon_0]$.

Gentzen points out that it is difficult to estimate or foresee how far one could advance in the second number class with purely number-theoretic methods. But he considers it to be beyond doubt that one manages to reach far beyond ε_0. Every segment of the second number class thus obtained thereby merges into a domain that must be assigned to pure number theory in a natural way. This thought appears evident, if one maps the ordinal numbers on the natural numbers, as Hilbert and Bernays [30, §5.3, pp. 373 ff.] have done for the domain of numbers below ε_0. By way of this method, it is shown that transfinite induction for such a segment of the second number class is a mode of inference that belongs to pure number theory. We are therefore entitled to say that Gentzen's proof that $TI[\varepsilon_0]$ cannot be proved with the "remaining" number-theoretic modes of inference shows, from another point of view, the incompleteness of formal systems (rich enough to contain elementary number theory) which Gödel proved with respect to the formal unprovability (undecidability) of number-theoretic sentences.[16]

The question of whether $TI[\varepsilon_0]$ can be accepted as a finitist rule of inference is extensively discussed in [30], but unfortunately it is not given a clear-cut answer.[17]

---

first-order arithmetic proceeds within the bounds of finitistically acceptable methods. In [2, §6], he presents a stricter version of his proof of finiteness by establishing an upper bound for the number of Gesamtersetzungen which can be formed at all. The upper bound depends on certain constants of the proof figure. In the light of the methodological requirements for a consistency proof, the dependence must be expressible via recursively defined functions. Yet it is not imperative to abandon the domain of the natural numbers when such recursive functions are introduced, since the natural numbers can be ordered according to every order type of the second number class.

[16] On Gentzen see also [71].

[17] In the Introduction to [30] (the Introduction was written in 1939), Bernays observes that Ack-

However, to all appearances Hilbert and Bernays come close to regarding $TI[\varepsilon_0]$ as a mode of inference that can be applied within the bounds of finitism. We are told (p. 387) that the more comprehensive the formalized theory under consideration is, the higher are the forms of the generalized induction principle that must be employed in the consistency proof. The methodic requirements for the contentual proof of a higher induction principle are said to provide the criterion for deciding which kind of methodic assumptions must be accepted as a basis for the "contentual attitude", if the consistency proof for the formalized theory is to be successful. Two possible interpretations may come to mind. First, whenever in $PA + TI[\varepsilon_0]$ the consistency of an important mathematical theory is provable, $TI[\varepsilon_0]$ can be considered to be a finistically permissible mode of inference. Second, in order to be able to carry out a consistency proof for a relatively strong theory such as second-order arithmetic, the use of a higher form of transfinite induction is requisite. Whether the latter is finitistically admissible, is of no real concern. No matter which interpretation was intended, Hilbert's original finitist point of view has been abandoned.

Let me briefly summarize. In [30], the mark of distinction that finitist metamathematics was supposed to wear on its sleeve in Hilbert's approach in the 1920s, which includes representability in intuition, surveyability, intuitive evidence and incontestable soundness, has largely disappeared in formalized metamathematics. In particular, the intuitive, down-to-earth modes of inference are now replaced with the loftier ones of a formalized metatheory. The new finitist parameters are kept flexible to make allowances for special proof-theoretic needs. Obviously, Hilbert and Bernays cannot have their cake and eat it, that is, they cannot make metamathematics possibly much stronger than PA by adding $TI[\varepsilon_0]$ or perhaps even more transfinite machinery to it and, in the same breath, defend a version of finitism which is still reasonably faithful to Hilbert's original approach. Their apparent attempt to do so

---

ermann is working out a modification of the consistency proof that he gave in [1]. The new modified proof is said to be valid for the full number-theoretic formalism. As we have seen, the modification concerns the use of transfinite induction. Bernays claims that, if Ackermann's new consistency proof were successful, Hilbert's original proof-theoretic approach should be seen as "rehabilitated" as far as its effectiveness is concerned. He adds that thanks to Gentzen's consistency proof one could hold that the "temporary fiasco" of proof theory was merely caused by the "overstretched" ("überspannten") methodic constraints imposed on metamathematics. (Bernays had already mentioned in [3] that the proof method applied by Gentzen [12] meets the fundamental requirements of Hilbert's (original) finitist point of view.) In my judgement, Bernays is mistaken here. The constraints that Hilbert had imposed on strict finitism in the 1920s were by no means "overstretched". If formalized metamathematics embraces transfinite induction, then Hilbert's original finitist point of view is completely abandoned and it does not make sense to assess his finitist approach of the 1920s from the viewpoint of a formalized metamathematical theory that has been extended to the transfinite. Hence, Gentzen's (1936 and 1938) and Ackermann's (1940) consistency proofs cannot be regarded as "rehabilitating" Hilbert's original programme.

in several places of their work is after all an exercise in futility.[18]

3 Hilbert's finitist proof theory and Gödel's second incompleteness theorem

In this final section, I wish to discuss the question of whether, contrary to a widely accepted view among Hilbert scholars and proof theorists, Hilbert's original programme of the 1920s was not refuted by Gödel's second incompleteness theorem. This does not mean that I sympathize with the idea of resuscitating this programme. I certainly do not believe that the programme can be salvaged *tout court*. Yet I do think that analyzing the question raised above is nonetheless of special interest for an adequate understanding of Hilbert's conception of metamathematics in the 1920s and beyond.[19]

For the sake of argument and in order to facilitate my exposition, I shall assume that Hilbert's finitist metamathematics of the 1920s (let us refer to it as "M") is clad in the garb of a formal theory (let us call it "MM"). MM results from M by couching M in a formal language. Note that MM need not be recursively enumerable. The question of whether MM is recursively enumerable, that is, axiomatizable, concerns only the complexity of the set of Gödel numbers of sentences belonging to MM within

---

[18] In his influential paper 'Partial Realizations of Hilbert's Program' [61], S. G. Simpson argues that Gödel's second incompleteness theorem does not rule out significant partial realizations of Hilbert's programme. Simpson regards Hilbert's programme as consisting of three steps. The first is to isolate the finitist part of mathematics; the second step is to reconstitute infinitistic mathematics as a fairly strong formal theory, Z_2, for example. The third step is to carry out finitist consistency proofs for that theory. Since Simpson endorses Tait's thesis (in [64]) that finitist reasoning is essentially primitive recursive in the sense of Skolem, the final step in Hilbert's programme is then to show that the consistency of Z_2 can be proved in the formal system PRA. If this could be achieved, then Z_2 would be conservative over PRA with respect to Π_1^0-sentences. Here is my brief assessment of Simpson's approach. However valuable and fruitful it may appear in its own right, it does not square with Hilbert's original programme in two respects. First, I disagree with Tait and Simpson that we can safely assume that the core of Hilbert's finitist metamathematics of the 1920s is adequately captured by PRA. Second, even if we granted that a formal version of Hilbert's finitist metamathematics of the 1920s is equal to PRA, partial realizations of Hilbert's programme would still be far removed from Hilbert's original conception of carrying out finitist consistency proofs for formalized mathematical theories. From the fact that a recursively enumerable theory T is Π_1^0-conservative over PRA, it follows precisely that the consistency of T cannot be proved in PRA. Admittedly, a proof of the Π_1^0-conservativeness of a theory T whose consistency does not appear to be secured may reinforce our confidence that T is indeed consistent. Nevertheless, such a relative consistency proof is a matter quite distinct from the direct consistency proofs that Hilbert sketched or merely envisaged in his essays on proof theory in the 1920s.

[19] Regarding a generalization of Gödel's incompleteness theorems for arithmetically definable theories see [37]; see especially the discussion of the second incompleteness theorem in §5.

the arithmetical hierarchy. It asks whether we *could* give a recursive set of axioms for *MM*. Of course, Hilbert does not and could not, at his time, employ the terminology of the arithmetical hierarchy. Roughly speaking, the arithmetical hierarchy is the sequence of Σ_m^0-formulae or of the sets defined through Σ_m^0-formulae. The latter are usually inductively defined.[20]

To begin with, since mathematics in its entirety embraces at least ZF, a "strong" version of Hilbert's programme would imply that the consistency of ZF is provable in PA. By contrast, if we consider a "weak" version of this programme, the requirement that PA proves its own consistency is usually taken to be a *conditio sine qua non* for the viability of the programme. One might be tempted to assume that in the light of Gödel's second incompleteness theorem the apparent unsatisfiability of this condition definitely refutes Hilbert's original programme. However, due to the fact that there are several ways of explicating the statement "PA proves the consistency of PA" it seems to me that this assumption is not well-grounded. The statement "PA proves the consistency of PA" may well be true if it is not understood in the ordinary way.

To see this, I suggest that we consider the following arithmetization of "PA is consistent": Let the representation pa' of a set of axioms of PA be a subformula of an arithmetical formula $Con_{pa'}^*$. We can define $Con_{pa'}^*$ to be an extensionally adequate arithmetization of "PA is consistent" if and only if the following condition obtains: PA is consistent $\Leftrightarrow \mathcal{N} \vdash Con_{pa'}^*$.[21] It then follows by definition that if PA is consistent, then "$\forall x(pa(x) \vee \neg pa(x))$" is an extensionally adequate arithmetization of "PA is consistent". However, it seems that putting it in this way does not reveal anything substantial about consistency. We should therefore try to strengthen the conditions under which an arithmetization of the metatheoretic consistency claim can be regarded as satisfactory.

"$Con_{pa'}^*$" is a numeralwise adequate arithmetization of "PA is consistent". "$Con_{pa'}^*$" is defined by "$\neg \exists x \text{Proof}_{pa'}^*(x, \bot)$", where pa' numerates PA in PA, and "$\text{Proof}_{pa'}^*(x, y)$" numerates the relation of PA-proof in PA, that is $\forall n, \varphi (n$ is a proof for φ in PA $\Leftrightarrow$ PA $\vdash \text{Proof}^*_{pa'}(n, \varphi))$ holds. Clearly, every numeralwise adequate arithmetization $Con_{pa'}^*$ of "PA is consistent" is extensionally adequate. J. B. Rosser [53], S. Feferman [9] and other logicians have shown that there are numeralwise adequate arithmetizations of the statement "PA is consistent" which, despite appearances, are provable in PA (if PA is consistent).

As to the unusual formalizations of "x is a proof for y in T", let me mention that

---

[20]For a detailed account of the arithmetical hiercharchy, cf. [62] and [36]. According to a theorem of W. Craig [6], there is for every axiomatizable theory T a recursive set of sentences A (a set of axioms or an axiomatization of T) such that T is equal to the deductive closure of A.

[21]"$\mathcal{N}$" refers here for the standard model of arithmetic.

in Rosser's version, for example, the predicate "proof of" appears to be modified,[22] while in Feferman's version, for example, we have an unusual representation τ of T. Suppose that we require for recursively enumerable extensions T of Q (Q = Robinson Arithmetic[23])

(a) that the arithmetization "$\text{Proof}_\tau(x,y)$" depends only on our choice of the representation τ of T, and

(b) that τ is in Π_0^0.

Then each consistency statement Con_τ, defined as "$\exists x \text{Proof}_\tau(x, \bot)$", is not provable in T, even if τ is not the ordinary representation of T and, hence, even if "$\text{Proof}_\tau(x,y)$" is not the ordinary formalization of the proof predicate for T.

Gödel's second incompleteness theorem shows merely for one special (though quite natural) numeralwise adequate arithmetization of "PA is consistent" its independence from PA. However, if the metamathematician has to prove $\text{Con}^*_{pa'}$ with finitist means, where $\text{Con}^*_{pa'}$ is any numeralwise adequate arithmetization of "PA is consistent", then, under a suitable interpretation of "finitist", he might be able to do this.

The upshot so far is this: One way of calling into question the view that Gödel's second incompleteness theorem is incompatible with Hilbert's original proof-theoretic programme consists in exploiting ambiguities in the formalization of the metatheoretical consistency statement. One can safely assume that in the course of developing his proof theory in the 1920s Hilbert was not aware of the possibility of formulating "unnatural" provability predicates in the sense just hinted at. And there is no reason to believe that he would have accepted the finitist proof of a clearly "unnatural" formalization of "ZF is consistent" as an adequate consistency proof for ZF. Be that as it may, certain remarks scattered throughout Hilbert's classical papers on proof theory suggest that he construed consistency proofs in a way which likewise differs significantly from the standard modern conception of consistency proofs. A standard consistency statement says that for all x, S proves that it is not the case that x is a proof for *falsum*. In [46], the non-standard notion of an *approximative* consistency proof was introduced and defined as follows (for axiomatizable theories S and T and representation τ of T):

S proves the approximative consistency of T if and only if $\forall n\ S \vdash \neg \text{Proof}_\tau(n, \bot)$

---

[22] For a theory with representation τ, Rosser [53] defines the provability predicate as follows: $\Delta_\tau(x) := \exists z(\text{Proof}_\tau(z,x) \wedge \neg \exists y_{\leq z}(\text{Proof}_\tau(y, \neg \ulcorner x) \vee \exists \xi_{\leq x}(x = \neg \ulcorner \xi \wedge \text{Proof}_\tau(y, \xi))))$. For consistent theories T with representation τ, Rosser's provability predicate is co-extensional with that used by Gödel.

[23] Q is a finitely axiomatized fragment of PA. Q is incomplete and undecidable.

The variable n ranges here over natural numbers which function as Gödel numbers of formulae or sequences of formulae of the language L_T. The definition reads as follows: The theory S proves the approximative consistency of T if and only if for all natural numbers n the sentence "$\neg\text{Proof}_\tau(n, \bot)$" is provable in S. I assume that the formalized proof predicate "$\text{Proof}_\tau(x, y)$" is the standard one. It is worth noting that we need not construe "τ" as a natural representation of a set of axioms of T. We must only make sure that "τ" is a Π_0^0-sentence, that is a sentence which is quantifier-free or contains at most bounded quantifiers. If we do this, then we are entitled to assume that "τ" can indeed be formulated in the language of MM.[24]

Whereas "Con_{pa}" is not PA-provable (assuming that PA is consistent), we can prove the approximative consistency of PA in PA, even the approximative consistency of ZF in Q (if ZF is consistent). Furthermore, it seems that approximative consistency proofs are in the spirit of Hilbert's finitist point of view of the 1920s, that is, that Hilbert could agree that in order to show finitistically the consistency of a formalized mathematical theory T it suffices to show finitistically its approximative consistency. Thus, it seems that for the restricted case in which the consistency of PA ought to be provable in a finitist fashion in PA, Hilbert's programme could be carried out.

In [23, p. 39], Hilbert describes the key idea of how a consistency proof for a set of axioms gets off the ground, as follows:

> We assume that we are given a concrete proof with the end-formula $0 \neq 0$; the presence of a contradiction can in fact be reduced to this case. Then, by considering the matter in a finitist and contentual way, we show that this cannot be a proof which meets our requirements.

This tallies with Hilbert's characterization of consistency proofs in [26, pp. 19 f.]. The preferred proof-strategy consists of two basic steps:

1. The starting point is the assumption that we are given a formalized proof qua concrete proof figure of an axiomatic mathematical theory T with the end formula "$0 \neq 0$" (or "$1 \neq 1$").

2. In a second step, it is shown by means of contentual finitist reasoning that "$0 \neq 0$" cannot be obtained as the end formula of a proof that starts from the axioms of T and proceeds by applying a finite number of times the formal inference rules of T, hence that "$0 \neq 0$" is not a provable T-formula.

---

[24] In [31, pp. 331 f.], Ignjatovic introduces the notion of an "almost finitistic consistency proof" which seems to amount to a natural formalization of an approximative consistency proof.

As Hilbert puts it succinctly in [24, p. 97]: "The point for us is to show that it is impossible to exhibit a proof of a certain kind."[25] Thus, a metamathematical consistency proof is designed to prove that in T it is *impossible* to carry out a proof of a well-specified nature, namely a proof whose end-fomula is "$0 \neq 0$".[26] In order to gain a more specific idea of this proof strategy and to see that its core statement is at least in close vicinity to an approximative consistency proof, let me summarize the main points that Hilbert makes in [23] in this respect. In doing so, I shall also refer to a consistency proof in [29] which is immediately reminiscent of the 1923 proof sketch.

In [23], Hilbert introduces four axiom groups which comprise ten quantifier-free axioms plus axiom V, 11: $A(\tau A) \to A(a)$, which is raised to the lofty status of the primordial and supreme transfinite axiom reigning over all other transfinite principles and axioms.[27] In *ordinary* language, the axiom states that if a predicate A applies to the object $\tau(A)$, then it applies to all objects a. In order to obtain V, 11, Hilbert introduces a logical function $\tau(A)$ or $\tau_a(A(a))$ which assigns a definite object $\tau(A)$ to each predicate $A(a)$. The function τ is to satisfy Axiom V, 11. The introduction of this operator on predicates rests on the idea that underlies Zermelo's axiom of choice. The axiom system is completed by four axioms that define the universal and existential quantifiers and are straightforward descendants of V, 11 (cf. [23], p. 38, [24], p. 96). Their addition to V, 11 is said to guarantee that

---

[25] To be sure, the statement that "$0 \neq 0$" is not a provable T-formula cannot be proved formally, since a formalized theory T does not contain such a statement.

[26] A natural explication of the consistency of a theory can be stated as follows: T is consistent if there is at least one unprovable formula in T. The alternative explication "T is consistent if it is impossible that every formula of T is provable in T" is slightly weaker. The first explication requires the actual exhibition of an unprovable T-formula, whereas the second rests content with a "proof of impossibility". In [29, pp. 219 f.], the authors observe that in order to prove the consistency of the axiom system (let us call it "AS"):

(J1)	$a = a$	(P_1)	$a' \neq 0$	($<_1$)	$a < a$
(J2)	$a = b \to (A(a) \to (A(b))$	(P_2)	$a' = b' \to a = b$	($<_2$)	$a < b \land b < c \to a < c$
				($<_3$)	$a < a'$

it suffices to recognize one formula of this formalism as unprovable, or in other words: to recognize as impossible that $0 \neq 0$ is proved from the axioms of AS.

[27] In [24], the fundamental transfinite axiom from [23] appears in the guise of the so-called logical ε-axiom

$$A(a) \to A(\varepsilon A).$$

This modification rests on the introduction of the choice function $\varepsilon_x A(x)$ or $\varepsilon(A)$ which replaces the old function $\tau(f)$. In [46], it is argued that in his metamathematics of the 1920's Hilbert tacitly makes assumptions of infinity, contrary to the strictly finitist constraints that he imposes on metamathematics. For a systematic investigation of assumptions of infinity in a (mathematical) theory T and of their strengths and weaknesses see [45].

all purely logical transfinite principles such as the so called Aristotelian principle $\forall a(A(a) \to Aa)$ (the first of six such principles listed by Hilbert) or the so called existential principle $A(a) \to \exists a(Aa)$ (the second) turn out to be provable formulae.

The consistency proof that Hilbert outlines for the axiom system (I) – (V) can be construed as consisting of two successive "partial" proofs. The first proof is concerned with the consistency of the ten axioms of groups (I) – (IV), the second with the consistency of V, 11 which Hilbert eventually turns into its twin companion V, 12. By taking the concrete proof (A) with the end-formula "$0 \neq 0$" as the jumping-off-point along the lines of an approximative consistency proof, the ensuing application of the proof strategy relies on a successive modification of (A) according to the following operations:

1. By way of reiterating and omitting certain formulae, (A) is dissected into what Hilbert calls "threads" (in [29, pp. 220 ff.] labelled "proof-threads" and explained there in much detail) which, starting from the axioms, eventually merge into the end-formula.

2. The variables that occur in (A) can be eliminated.

3. (A) can be rearranged in such a way that every formula contains only numerical signs besides the logical signs.

4. Every formula is presented in such a way that it appears in the guise of a certain logical "normal form".

Once the operations (1) – (4) have been carried out on the initially given proof (A), we can determine for every formula of its modified version (A*) whether it is true or false. "Now, if the proof so presented is to meet all our requirements, then each formula of the proof would, as it turns out, have to pass successively this test" ([23], p. 39). Consequently, the consistency of the axiom-groups I – IV would be established.

As far as the proof of the consistency of Axiom V, 11 is concerned, it must first be noted that the logical function τ assigns a number to every *mathematical* function f. Thus, τ gains the status of a whole-numbered function of functions, such that if f is a certain function, τ is a certain number. Axiom V, 11 is then converted into Axiom V, 12: $f(\tau(f)) = 0 \to f(a) = 0$. The strategy that Hilbert pursues in order to prove "the consistency of the transfinite function $\tau(f)$" is an extension of the prior consistency proof for the Axioms 1 to 10 by transferring it to the present case and by carrying out again a number of operations. The details need not be fleshed out here. Suffice it to mention that in the end all the formulae that derive from Axiom V, 12 have indiscriminately the form $\varphi(z) = 0 \to \varphi(s) = 0$, "and these formulae are correct in themselves, since the formula preceding the implication-sign is false. The proof once again becomes a proof with purely numerical formulae which are correct, so that the end-formula cannot be $0 \neq 0$" [23, p. 41].

So much for [23]. To my mind, it seems to suggest itself that in essence the outlined proof is intended to establish the consistency of a mathematical theory T in the fashion of an approximative consistency proof. For a given concrete proof figure F or for every given concrete proof figure F, it has to be shown that F is not a proof for *falsum* in T.[28]

A few concluding remarks on approximative consistency proofs are in order. If we turn M via formalization into MM, in particular, if we replace proofs qua concrete and surveyable figures with proofs qua numbers, and contentual metamathematical considerations with proofs in MM, we seem to have the choice between three ways of formally characterizing the salient feature of Hilbert's conception of finitist metamathematical consistency proofs in the 1920s:

(a) $\forall n$(If n is a proof for $\bot$ in S, then $MM \vdash$ 'n is no proof in σ').[29]

(b) $\forall n \forall \Phi$(If n is a proof for Φ in S, then $MM \vdash \Phi \neq \bot$').

(c) $\forall n MM \vdash \neg \text{Proof}_\sigma(n, \bot)$.

However, we need not give preference either to (a) or to (b) or to (c). The reason is that we can prove that (a) is equivalent to (b), that (b) is equivalent to (c) and, hence, that (a) is equivalent to (c), *if* two conditions are satisfied: (a) S must be recursively enumerable with Π_0^0-representation σ, and (b) MM must decide every Π_0^0-sentence.

In sum, I think that by appreciating the distinctive character of the notion of an approximative consistency proof we can make good sense of the nature of the consistency proofs that Hilbert outlines both in his classical papers on proof theory in the 1920s and in [29]. Unsurprisingly, in at least one important respect an

---

[28] As I mentioned before, in [29] the authors carry out a consistency proof for the axiom system AS (cf. pp. 219). The proof is informal and follows grosso modo the pattern of the 1923 proof sketch. Yet it is not entirely clear whether Hilbert and Bernays intend to show metamathematically "There is no proof in AS for falsum" or only for every concretely given proof figure F that F is not a proof for falsum in AS. The beginning of the proof speaks in favour of the second option. I assume therefore that Hilbert and Bernays carry out what is or at least comes fairly close to an informal version of an approximative consistency proof. "We now imagine that we are given such a proof figure with the end formula $0 \neq 0$. On this proof figure two processes can be effected one after another which we call dissolution of the proof figure into 'proof-threads' and elimination of the free variables" ([29], p. 220; cf. p. 298).

[29] "n is no proof in σ' is to be regarded here as a formalization of 'n is no proof in S', which is Π_0^0 if σ is Π_0^0.

approximative consistency proof is methodologically not on a par with a standard consistency proof. If we wish to establish the consistency of a mathematical theory T by carrying out a standard consistency proof for T in MM, we must only assume that MM is sound. By contrast, if we intend to establish the consistency of T via an approximative consistency proof for T in MM, we must presuppose not only the soundness of MM but also the consistency of T. Thus, even if we succeeded in advancing a strong argument for the soundness of MM, what we accomplish in the case of an approximative consistency proof is only that we prove the consistency of T relative to the assumption that T is consistent. This is perhaps one of the reasons why modern proof theorists prefer to carry out a consistency proof in the standard way and not along the lines of an approximative consistency proof.

Finally, let me make a remark on the scope of approximative consistency proofs which was already apparent in the preceding account. MM can be used to prove the approximative consistency of a mathematical theory T only if T has approximately the same "strength" as MM. Owing to this circumstance and due to the "relative weakness" of MM, assuming that one could succeed in giving a consistency proof for, say, second-order arithmetic in MM is simply an illusion. Thus, Hilbert's programme of the 1920s, which was designed to furnish the resources for carrying out a finitist consistency proof at least for second-order arithmetic, falls short even if such a proof is explicitly characterized as an approximative consistency proof in terms of the definition of this notion given above. However, in principle it can be said that the proof-theorist, who follows strict finitist guidelines à la Hilbert and, hence, does not license the formulation of genuine Π_1^0-sentences in the language of metamathematics, can still reasonably make consistency statements in accordance with the definition of an approximative consistency proof and eventually establish the consistency of a formalized mathematical theory T, provided that the strength of T does not exceed that of MM.

The foregoing considerations suggest that a "weak" version of Hilbert's programme can be seen to be in tune with Gödel's second incompleteness theorem, if "PA proves the consistency of PA" is considered to be tantamount to "PA proves the approximative consistency of PA". However, the other side of the coin is that the compatibility is achieved at the expense of modifying standard consistency proofs in a way which most modern proof theorists would probably regard as unsatisfactory. Seen in this way, it seems more appropriate to show that Hilbert's metamathematical approach can be considered to be compatible with Gödel's second incompleteness theorem by using only what are clearly *natural* provability predicates. Admittedly, we are facing difficulties when we attempt to provide a materially adequate, generally applicable and purely mathematical definition of "natural proof predicate" or "natural provability predicate". However, when we consider mathematical theories such as PA, Z_2 or ZF, we have a clear idea about which codings of the relevant metatheoretical provability predicates are natural. To arrive at this idea it is not mandatory to provide a general mathematical explication of what counts as a natural provability predicate. It is therefore appropriate to consider theories T with which we associate a clear idea about what counts as a natural representation τ of T.

Suppose that we can show the soundness of a mathematical theory T in M, not just the consistency of T. Since M is, from Hilbert's finitist point of view, indisputably sound, a soundness proof for T in MM implies that T is sound. Thus, it seems that we can use T to carry out convincing consistency proofs for theories S. For it is by simple virtue of the metamathematically and finitistically established soundness of T that we know that S *is* consistent if T proves the consistency of S.[30]

In rough outline, this idea can be developed as follows.[31] Let us consider a formalization of the soundness statement above, the local reflection principle for $\Pr_\tau$:

$$\text{For all sentences } \Psi \in L_T,\ MM \vdash \Pr_\tau(\Psi) \to \Psi \qquad (+)$$

τ is here a representation of T, and $\Pr_\tau$ is a natural provability predicate of T. By virtue of (+) and the assumption that M as well as MM contain only true sentences, "$\Pr_\tau(\Psi) \to \Psi$" is true for all sentences Ψ that is, if T proves Ψ then Ψ is true (in $\mathcal{N}$).

If in addition to (+) T proves Con_σ, then we know that S is consistent. This gives rise to the following definition:

Definition. *Let U and S be theories, and let S be recursively enumerable with Σ_1^0-representation σ; U proves the consistency of S indirectly in one step just in case* $\exists T(T$ *is a theory with representation τ and $T \vdash \text{Con}_\sigma$ and $\forall\Psi(\Psi \in \Pi_1^0 \Rightarrow U \vdash \Pr_\tau(\Psi) \to \Psi))$.*

---

[30] I am aware that in his papers of the 1920s Hilbert did not mention this possibility. One possible reason for not mentioning it is that those soundness statements, which are not equivalent to consistency statements, are of greater complexity than Π_1^0-sentences and, hence, cannot be formulated in the language of finitist metamathematics. However, I assume that M or MM appears in the formal guise of PA. Under this assumption, we can formalize soundness statements in the formal language of MM by using formalized provability predicates and partial truth predicates. Regarding the notion of partial truth-predicates see [43, §1.1.8, pp. 16 ff.]. "$Tr_{\Sigma_2^0}$", for example, is a partial truth-predicate. It refers to the set of Σ_2^0-sentences which are true in the standard model $\mathcal{N}$. Note that the theory of partial truth-predicates can be developed in the theory (QF-IA) (QF = quantifier-free; IA = induction axiom). (QF-IA) is the conservative extension of PRA obtained by adding first-order logic to PRA.

[31] The ensuing discussion draws on [47]. Thanks to the editor of *Dialectica*, Philipp Blum, for granting permission to reuse this material in a slightly modified form.

This definition is acceptable for the proof-theorist endorsing Hilbert's finitism, provided that he or she can state an example of such a theory T (call it "interpolant"). To be sure, it is not required that the interpolant T is a finitist theory. The adequacy of the *definiendum* can be shown by proving the following Lemma:

Lemma 1. *If MM proves the consistency of S indirectly in one step, then S is consistent.*

The proof is straightforward.
Since on assumption $T \vdash \mathrm{Con}_\sigma$ holds,

$$\text{``}\mathrm{Pr}_\tau(\mathrm{Con}_\sigma)\text{'' is true.} \tag{1}$$

Since on assumption we have for all $\Psi \in \Pi_1^0$ $MM \vdash \mathrm{Pr}_\tau(\Psi) \to \Psi$ and Con_σ is Π_1^0, it holds that $MM \vdash \mathrm{Pr}_\tau(\mathrm{Con}_\sigma) \to \mathrm{Con}_\sigma$. Hence,

$$\text{``}\mathrm{Pr}_\tau(\mathrm{Con}_\sigma) \to \mathrm{Con}_\sigma\text{'' is true.} \tag{2}$$

From (1) and (2) follows that "Con_σ" is true.[32]

The Lemma shows that in order to prove the consistency of PA in a finitistically acceptable fashion, it would likewise suffice to prove the consistency of PA indirectly in one step in MM. Yet in the light of considerations closely linked to Gödel's incompleteness theorems, it seems that this strategy is bound to fail. If it were feasible, there would have to exist an interpolant T such that $MM \vdash \mathrm{Con}_\tau$ and $T \vdash \mathrm{Con}_{pa}$. Assuming here that MM is a subtheory of PA, we obtain also $PA \vdash \mathrm{Con}_\tau$. Hence, PA would be "stronger" than T, and, conversely, T would be "stronger" than PA. Thus, we seem to face a no-go situation, because the stronger-than-relation employed here is transitive and irreflexive.

In short, it is tempting to believe that at least for recursively enumerable extensions S of (QF-IA), Gödel's second incompleteness theorem rules out that we can give indirect consistency proofs for S in S by using natural provability predicates. However, it seems to me that this belief is ungrounded. Our task is then to prove a theorem which expresses that recursively enumerable extensions of (QF-IA) prove their own consistency indirectly in one step. Admittedly, when we deal with theories like (QF-IA), which prove only very simple instances of the induction scheme, the central idea of the proof does not emerge clearly. It is therefore useful to prove first the corresponding theorem for PA.

---

[32]Since for recursively enumerable S with Σ_1^0-representation σ, Con_σ is in Π_1^0, the MM-provability of the consistency statement in the proposed definition above suffices for proving the Lemma.

Theorem 1. *PA proves its own consistency indirectly in one step.*

The proof of Theorem 1 is straightforward:

We can choose (QF-IA) + $Tr_{\Pi_1^0}$ for the interpolant T, with "natural" Π_1^0-representation τ: $\tau(x) \equiv (\text{QF-IA})(x) \vee Tr_{\Pi_1^0}(x)$.

Suppose that PA is consistent — if it were not consistent, PA would prove "every statement", including Con_{pa}.

$$T \vdash \text{Con}_{pa} \tag{I1}$$

and

$$\text{PA} \vdash \text{RFN}[\tau]. \tag{I2}$$

already yield Theorem 1. It is clear that $T \vdash \text{Con}_{pa}$; for on assumption, "Con_{pa}" is in $Tr_{\Pi_1^0}$.

Moreover, in order to establish (I2) we only need to show

$$\text{PA} \vdash \text{RFN}[\tau]. \tag{*}$$

(*) follows immediately from (a) the fact that PA proves the uniform reflection principle for (QF-IA) and from (b) a result which may be called the "Extension Lemma":

Lemma 2 (Extension Lemma). *Let S be an arithmetically definable theory with representation σ and $k \geq 1$. Then $\text{PA} + \text{RFN}[\sigma] \vdash \text{RFN}[\sigma + Tr_{\Sigma_k^0}]$.*

Theorem 2. *If S is a recursively enumerable extension of (QF-IA), then S proves its own consistency indirectly in one step.*

To be sure, the formalizations of the metatheoretical consistency statements that occur in Theorem 1 and Theorem 2 are the natural ones.[33]

I conclude by adding one more brushstroke to the picture that I have painted of approximative and indirect indirect consistency proofs.

First, T satisfies suitable forms of Löb's derivability conditions [40] and, consequently, Gödel's incompleteness theorems. T and PA thus show that there are sound theories to which these theorems apply, but which nonetheless mutually prove their consistency.[34]

---

[33] For the proofs of the Extension Lemma and Theorem 2 see [47]. In [44], Niebergall analyzes theories which are not recursively enumerable and prove their own consistency. Those theories are not subject to the dictates of Gödel's incompleteness theorems.

[34] See [43, §3.3] where mutual consistency proofs are investigated in some detail. Niebergall first gives a simple example of correct Gödel-style theories S and T which mutually prove their consistency. In what follows, he argues, by choosing as an example two axiomatizable theories,

Second, for arithmetically definable sound theories T — extending (QF-IA) — which are complete for their complexity class, we obtain the same result as for PA. T proves its own consistency indirectly in one step, athough it is not assumed that T is recursively enumerable or a subtheory of PA.

Third, let S, T be arithmetically definable theories with representations σ and τ such that they satisfy both Löb's derivability conditions and Gödel's incompleteness theorems. If we now define $S \, \sigma \, T$ such that $T \vdash \text{Con}_\sigma$, it seems that ρ is a strict partial ordering which can be construed as a stronger-than-relation between theories. This applies in fact to axiomatizable theories S, T with Π_0^0-representations σ, τ, extending (QF-IA). The matter stands differently for non-axiomatizable theories. It follows from Theorem 1 that in this case ρ is not even transitive. Here, we have a clear example of the circumstance that there are stronger-than-relations familiar from axiomatizable theories which are rendered inadequate as soon as we deal with non-axiomatizable theories.

One concluding remark. Even if Gödel's Second Incompleteness Theorem were false, and thus even if, say, PA were to prove its own consistency or soundness, we would still not know that PA is consistent or sound. No matter whether we prove in a theory S the consistency of a theory T, the approximative consistency of T or almost finitistically the consistency of T, we are bound to presuppose the soundness of S in order to make the claim that T *is* consistent appear well-founded. However, it seems to me that consistency proofs for formalized mathematical theories T must be carried out in a theory in whose soundness we have strong, though *informal* reasons to believe. Whenever such reasons could be presented and underpinned with a substantial argument, a consistency proof might turn out as a blessing in disguise for the proof theorist.

4 Acknowledgements

Parts of this essay were presented at University of Chicago; University of Munich; Keio University (Tokyo); Kyoto University; Tohoku University (Sendai); Japan Advanced Institute of Science and Technology (Nomi); Tokyo Metropolitan University; Hokkaido University (Sapporo); Nanyang Technological University (Singapore); Fudan University (Shanghai); Munich Center for Mathematical Philosophy; University of Vienna (Institute Vienna Circle, Philosophy Lectures); Vrije Universiteit Brussel

---

for the impossibility of mutual consistency proofs. Based on this, he argues that "sufficiently strong" theories, which share their degree of complexity, cannot mutually prove their consistency. Finally, Niebergall shows by appeal to theories of distinct complexity, which are complete for their complexity class, that mutual consistency proofs "frequently occur".

(Center for Logic and Philosophy of Science); Université Paris Diderot, Séminaire de philosophie des mathématiques; University of Salamanca; University of Naples Federico II (Department of Statistics and Mathematics); at a colloquium in Bayrischzell (Bavaria), Elite Master Course "Theoretical and Mathematical Physics" at Ludwig-Maximilians-Universität Munich (LMU) and Technische Universität Munich (TU) (invited principal guest speaker); Sociedad Argentina de Análisis Filosófico (Buenos Aires); International Conference 'The Emergence of Structuralism and Formalism', 24-26 June 2016, Charles University Prague, Czech Academy of Sciences (invited speaker); 'Proof Theory Afternoon' with Mario Piazza, Pierluigi Graziani, Ulrich Kohlenbach, Carlo Nicolai and Matthias Schirn, University Chieti-Pescara, 11 April 2017 (invited speaker).

Thanks to the audiences for interesting discussion. Special thanks are due to María Manzano, Yuko Murakami, Lina Stina Jansson, Friedrich Stadler, Ivan Smadja, David Svoboda, Alberto Moretti, Eduardo Barrio, Jean Paul van Bendegem, Patrick Allo, Michael Resnik, Michael Detlefsen, Michael Kremer, Jan von Plato, Andrei Rodin, Mario Piazza, Pierluigi Graziani, Ulrich Kohlenbach, Carlo Nicolai, Mitsuhiro Okada, Koji Nakatogawa, Satoshi Tojo, Katsuhiko Sano, Kazuyuki Nomoto, Kengo Okamoto and Eberhard Guhe. Last but not least, I thank Michael Gabbay for his advice, the conversion of my word-typescript into LaTeX and the preparation of the final version for publication in this journal and also for discussing matters of style.

References

[1] W. Ackermann. Begründung des "tertium non datur" mittels der Hilbertschen Theorie der Widerspruchsfreiheit. *Mathematische Annalen*, 93:1–36, 1925.

[2] W. Ackermann. Zur Widerspruchsfreiheit der Zahlentheorie. *Mathematische Annalen*, 117:162–194, 1940.

[3] P. Bernays. Hilbert's Untersuchungen über die Grundlagen der Arithmetik. In [27], pages 196–216. 1935.

[4] W. Buchholz. On Gentzen's first consistency proof for arithmetic. In [35], pages 63–87. 2015.

[5] G. Cantor. *Gesammelte Abhandlungen mathematischen und philosophischen Inhalts*. Berlin, 1932. Zermelo, E. (editor), reprint Georg Olms Verlagsbuchhandlung, Hildesheim 1966.

[6] W. Craig. On axiomatizability within a system. *The Journal of Symbolic Logic*, 18:30–32, 1953.

[7] M. Detlefsen. *Hilbert's Program. An Essay on Mathematical Instrumentalism*. Reidel, Dordrecht, 1986.

[8] W. B. Ewald and Sieg W., editors. *David Hilbert's Lectures on the Foundations of Arithmetic and Logic, 1917 to 1933.* Springer, Heidelberg, 2013.

[9] S. Feferman. Arithmetization of metamathematics in a general setting. *Fundamenta Mathematicae*, XLIX:35–92, 1960.

[10] S. Feferman. Hilbert's program relativized: Proof-theoretical and foundational reductions. *The Journal of Symbolic Logic*, 53:364–383, 1988.

[11] M. Ganea. Two (or three) notions of finitism. *The Review of Symbolic Logic*, 3:119–144, 2010.

[12] G. Gentzen. Die Widerspruchsfreiheit der reinen Zahlentheorie. *Mathematische Annalen*, 112:493–565, 1936.

[13] G. Gentzen. Neue Fassung des Widerspruchsfreiheitsbeweises für die reine Zahlentheorie. *Forschungen zur Logik und zur Grundlegung der exakten Wissenschaften*, 4:19–44, 1938.

[14] G. Gentzen. Beweisbarkeit und Unbeweisbarkeit von Anfangsfällen der transfiniten Induktion in der reinen Zahlentheorie. *Mathematische Annalen*, 119:140–161, 1943.

[15] G. Gentzen. *The Collected Papers of Gerhard Gentzen.* North-Holland, Amsterdam, 1969. Szabo, M. E. (editor).

[16] K. Gödel. Über formal unentscheidbare Sätze der Principia Mathematica und verwandter Systeme I. *Monatshefte für Mathematik und Physik*, pages 173–198, 1931.

[17] K. Gödel. Über eine bisher noch nicht benützte Erweiterung des finiten Standpunktes. *Dialectica*, 12:280–287, 1958. German original and English translation in [18], 240-251.

[18] K. Gödel. *Collected Works, vol. I, Publications 1929-1936.* Oxford University Press, Oxford, 1986. edited by S. Feferman, J. Dawson, S. C. Kleene, G. Moore, R. M. Solovay and J. van Heijenoort.

[19] J. Van Heijenoort. *From Frege to Gödel: A Source Book in Mathematical Logic, 1879–1931.* Harvard University Press, 1967.

[20] J. Herbrand. Sur la non-contradiction de l'arithmétique. *Journal für die reine und angewandte Mathematik*, 166:1–8, 1931. English translation in [19], pp. 618-628.

[21] D. Hilbert. *Grundlagen der Geometrie.* Teubner, Leipzig, 1st edition, 1899. 7th revised and enlarged edition 1930.

[22] D. Hilbert. Neubegründung der Mathematik: Erste Mitteilung. *Abhandlungen aus dem mathematischen Seminar der Hamburger Universität*, 1:157–177, 1922. Reprinted in [28, pp. 12-32].

[23] D. Hilbert. Die logischen Grundlagen der Mathematik. *Mathematische Annalen*, 88:151–165), 1923. Reprinted in [28, pp. 33-46] (and in [27, pp. 178-191]).

[24] D. Hilbert. Über das Unendliche. *Mathematische Annalen*, 95:161–190, 1926. Reprinted in [28], pp. 79-108.

[25] D. Hilbert. Die Grundlagen der Mathematik. *Abhandlungen aus dem Mathematischen Seminar der hamburgischen Universität*, 6:65–85, 1928. (Reprinted as appendix IX of [21], 289-312).

[26] D. Hilbert. Die Grundlegung der elementaren Zahlenlehre. *Mathematische Annalen*,

104:484–494, 1931. (Abbreviated version in [27], Vol. III, 192-195; complete English translation in [41], 266-273).

[27] D. Hilbert. *Gesammelte Abhandlungen*, volume III. Springer Verlag, Berlin, Heidelberg, New York, 1935,1970.

[28] D. Hilbert. *Hilbertiana. Fünf Aufsätze*. Wissenschaftliche Buchgesellschaft, Darmstadt, 1964.

[29] D. Hilbert and P. Bernays. *Grundlagen der Mathematik*, volume I. Springer Verlag, Berlin, Heidelberg, New York, 1st edition, 1934. second edition with modifications and supplementations, 1968.

[30] D. Hilbert and P. Bernays. *Grundlagen der Mathematik*, volume II. Springer Verlag, Berlin, Heidelberg, New York, 1st edition, 1939. 2nd edition with modifications and supplementations, 1970.

[31] A. Ignjatovic. Hilbert's program and the Omega-Rule. *The Journal of Symbolic Logic*, 59:322–343, 1994.

[32] L. Incurvati. On the concept of finitism. *Synthese*, 192:2413–2436, 2015.

[33] D. Isaacson. Some considerations on arithmetical truth and the ω-rule. In M. Detlefsen, editor, *Proof, Logic and Formalization*, pages 94–138. Routledge, London, 1992.

[34] R. Kahle. Gentzen's consistency proof in context. In [35], pages 3–24. 2015.

[35] R. Kahle and M. Rathjen, editors. *Gentzen's Centenary. The Quest for Consistency*. Springer, 2015.

[36] R. Kaye. *Models of Peano Arithmetic*. Oxford University Press, Oxford, 1991.

[37] M. Kikuchi and T. Kurahashi. Generalizations of Gödel's incompleteness theorems for Σ_n-definable theories. *The Review of Symbolic Logic*, 10:603–616, 2017.

[38] U. Kohlenbach. *Applied Proof Theory: Proof Interpretations and their Use in Mathematics*. Springer Monographs in Mathematics, Heidelberg, 2008.

[39] G. Kreisel. Hilbert's programme. *Dialectica*, 12:207–238, 1958. revised version in *Philosophy of Mathematics. Selected Readings*, editors P. Benacerraf & H. Putnam, Cambridge University Press, Cambridge 1964, second edition 1983, 207-238.

[40] M. Löb. Solution to a problem of Leon Henkin. *The Journal of Symbolic Logic*, 20:115–118, 1955.

[41] P. Mancosu. *From Brouwer to Hilbert. The Debates on the Foundations of Mathematics in the 1920s*. Oxford University Press, Oxford, 1998.

[42] P. Mancosu. Essay review of W. Ewald and W. Sieg, editors, *David Hilbert's Lectures on the Foundations of Mathematics*, Springer, 2013. *Philosophia Mathematica*, 23:126–135, 2015.

[43] K.-G. Niebergall. *Zur Metamathematik nichtaxiomatisierbarer Theorien*. University of Munich, CIS, 1996.

[44] K.-G. Niebergall. Natural representations and extensions of Gödel's second theorem. In M. Baaz and S. D. Friedman, editors, *Logic Colloquium '01*, pages 350–368. Association for Symbolic Logic, Urbana, IL and Wellesley, MA, 350-ÅŞ368, 2005.

[45] K.-G. Niebergall. Assumptions of infinity. In G. Link, editor, *Formalism and Beyond. On the Nature of Mathematical Discourse*, pages 229–274. Walter de Gruyter, Boston, Berlin, 2014.

[46] K.-G. Niebergall and M. Schirn. Hilbert's finitism and the notion of infinity. In [54], pages 271–305. 1998.

[47] K.-G. Niebergall and M. Schirn. Hilbert's programme and Gödel's theorems. *Dialectica*, 56:347–370, 2002.

[48] C. Parsons. Finitism and intuitive knowledge. In [54], pages 247–270.

[49] C. Parsons. *Mathematical Thought and Its Objects*. Cambridge University Press, Cambridge, 2008.

[50] V. Peckhaus. *Hilbertprogramm und kritische Philosophie. Das Göttinger Modell interdisziplinärer Zusammenarbeit zwischen Mathematik und Philosophie*. Vandenhoeck & Ruprecht, Göttingen, 1990.

[51] V. Peckhaus. Hilbert's axiomatic programme and philosophy. In E. Knobloch and D. E. Rowe, editors, *The History of Modern Mathematics, vol. 3: Images, Ideas, and Communities*, pages 91–112. Academic Press, Boston, 1994.

[52] D. Prawitz. A note on how to extend Gentzen's second consistency proof to a proof of normalization for first order arithmetic. In [35], pages 131–176.

[53] J. B. Rosser. Extensions of some theorems of Gödel and Church. *The Journal of Symbolic Logic*, 1:87–91, 1936.

[54] M. Schirn, editor. *The Philosophy of Mathematics Today*. Oxford University Press, Oxford, 1998.

[55] M. Schirn and K.-G. Niebergall. Extensions of the finitist point of view. *History and Philosophy of Logic*, 22:135–161, 2001.

[56] M. Schirn and K.-G. Niebergall. What finitism could not be. *Crítica*, 35:43–68, 2003.

[57] M. Schirn and K.-G. Niebergall. Finitism = PRA? on a thesis of W.W. Tait. *Reports on Mathematical Logic*, 39:3–24, 2005.

[58] W. Sieg. Hilbert's programs: 1917-1922. *Bulletin of Symbolic Logic*, 5:1–44, 03 1999. Reprinted in [60].

[59] W. Sieg. Hilbert's proof theory. In D. M. Gabbay and J. Woods, editors, *Handbook of the History of Logic*, pages 321–384. North-Holland, Amsterdam, Boston, 2009.

[60] W. Sieg. *Hilbert's Programs and Beyond*. Oxford University Press, Oxford, 2013.

[61] S. G. Simpson. Partial realizations of Hilbert's program. *The Journal of Symbolic Logic*, 53:349–363, 1988.

[62] C. Smorynski. The incompleteness theorems. In J. Barwise, editor, *Handbook of Mathematical Logic*, pages 821–865. North-Holland, Amsterdam, 1977.

[63] S. Stenlund. Different senses of finitude: An inquiry into Hilbert's finitism. *Synthese*, 185:335–363, 2012.

[64] W. W. Tait. Finitism. *The Journal of Philosophy*, 78:524–546, 1981.

[65] W. W. Tait. Remarks on finitism. In R. Sommer W. Sieg and C. Talcott, editors,

Reflections on the Foundations of Mathematics: Essays in Honor of Solomon Feferman. A. K. Peters and the Association for Symbolic Logic, Wellesley, MA, 2002.

[66] W.W. Tait. Gödel on intuition and Hilbert's finitism. In C. Parsons S. Feferman and S. Simpson, editors, *Kurt Gödel: Essays for His Centenial*, volume 33 of *Lecture Notes in Logic*, pages 88–108. Association for Symbolic Logic, 2010.

[67] W.W. Tait. Gentzen's original consistency proof and the bar theorem. In [35], pages 213–228. 2015.

[68] J. P. van Bendegem. A defense of strict finitism. *Constructivist Foundations*, 7(2):141–149, 2012.

[69] J. von Neumann. Zur Hilbertschen Beweistheorie. *Mathematische Zeitschrift*, 26:1–46, 1927.

[70] J. von Neumann. Selected letters. In M. Rédei, editor, *History of Mathematics*, volume 27. American Mathematical Society, London Mathematical Society, 2005.

[71] J. von Plato. Gentzen's logic. In D. M. Gabbay and J. Wood, editors, *Handbook of the History of Logic*, volume 5, pages 667–721. Elsevier North-Holland, Amsterdam, Boston, 2009.

[72] R. Zach. *Hilbert's Finitism: Historical Philosophical and Mathematical Perspectives.* PhD thesis, University of California, Berkeley, 2001.

[73] R. Zach. Hilbert's program then and now. In D. M. Gabbay and J. Wood, editors, *Handbook of the Philosophy of Science*, volume 5, pages 411–447. Elsevier North-Holland, Amsterdam, Boston, 2006.

[74] E. Zermelo. Beweis daß jede Menge wohlgeordnet werden kann. *Mathematische Annalen*, 59:514–516, 1904.

[75] E. Zermelo. Neuer Beweis, für die Möglichkeit einer Wohlordnung. *Mathematische Annalen*, 65:107–128, 1908.

Received June 2017

Formalism and Structuralism, a Synthesis: the Philosophical Ideas of H. B. Curry

Jonathan P. Seldin
Department of Mathematics and Computer Science, University of Lethbridge, Lethbridge, Alberta, Canada
jonathan.seldin@uleth.ca
http://www.cs.uleth.ca/~seldin

Abstract

The call for the conference "The Emergence of Structuralism and Formalism" contrasts *formalism*, which it describes as the manipulation of meaningless symbols, with *structuralism*, which it describes as treating mathematics as a subject with a subject matter.

H. B. Curry identified himself as a formalist, probably because of the influence of David Hilbert (under whose direction he earned his doctorate in 1928–29), but believed that all mathematical statements have a definite subject matter. What really distinguished Curry's views is that he thought that the only subject matter that mathematics has is generated by mathematics itself. This might lead to a view that we humans are programmed to engage in some mathematical activity, and this is where mathematics starts.

I believe that Curry is better considered to be a kind of structuralist than a formalist the way those words are used today.

Keywords: Formalism, Structuralism

The call for the conference "The Emergence of Structuralism and Formalism", at which this paper was presented, describes *formalism* as the manipulation of meaningless symbols. It contrasts this with *structuralism*, which it describes as treating mathematics as a subject with a subject matter.

H. B. Curry identified himself as a formalist in [2], which is a shortened form of the original manuscript of [3], where he first defined his idea of formalism, and most of which was written in 1939. This was early in his career, and these two works are often the only works by which Curry's philosophical ideas are known. Seek for example, [7, Chapter 6, §5], where [3] is mistakenly identified as a mature work.

But Curry did not believe that mathematics involves the manipulation of meaningless symbols. In fact, there was an argument against "meaningless concepts" in his dissertation, [1]. So Curry thought that mathematical statements have meaning; i.e., that mathematics has a subject matter. When he began to write about the philosophy of mathematics in 1939, Curry wrote in [3, p. 3]:

> There are three main types of opinion as to the nature of this subject matter, viz.: 1) realism, or the view that mathematical propositions are true insofar as they correspond with our physical environment; 2) idealism, which relates mathematics to mental objects of one sort or another; and 3) formalism.

Curry went on to say that realism was no longer taken seriously by most mathematicians, and under idealism he identified both Platonism and intuitionism (at least as understood by Brouwer and Heyting).

He then proposed to define mathematics as the "science of formal systems" [3, p. 56]. He goes on to say:

> This definition should be taken in a very general sense. The incompleteness theorems mentioned at the close of [the preceding chapter] show that it is hopeless to find a single formal system which will include all of mathematics as ordinarily understood. Moreover the arbitrary nature of the definitions which can constitute the primitive frame of a formal system shows that, in principle at least, all formal systems stand on a par. The essence of mathematics lies, therefore, not in any particular kind of formal system but in formal structure as such. The considerations of the preceding section show furthermore that we must include metapropositions as well as elementary ones. Indeed, all propositions having to do with one formal system or several or with formal systems in general are to be regarded as purely mathematical in so far as their criteria of truth depend on formal considerations alone, and not on extraneous matters.

Thus, for Curry, formalism meant that mathematics could be taken to be statements *about* formal systems, or, in other words, the metatheory of formal systems. In fact, Curry's notion of formal system was different from the usual one, and did not need to include the connectives and quantifiers of logic. Consider, for example, the following formal system for the natural numbers:[1]

---

[1] This is the system called $\mathcal{N}_1$ in [5], which, in turn, is the system of [4, p.256] and of [3, Example 1, page 18].

Example. The formal system $\mathcal{N}$ is defined as follows:

- Atomic term: 0

- Primitive (term forming) operation: forms $t|$ from t

- Terms: $0||\ldots|$

- Primitive predicate: $=$
  Elementary statements: $t_1 = t_2$

- Axiom: $0 = 0$

- Rule: $t_1 = t_2 \Rightarrow t_1| = t_2|$

Note that this is a really simple formal system: the theorems are precisely those elementary statements of the form $t = t$, where t is a term of the system.

In terms of Curry's ideas, most classical mathematics can be obtained as part of the metatheory of this formal system provided that one allows sufficiently strong methods of proof in the metatheory. Curry never expected it to be possible to obtain all mathematics in this way from one formal system, but he did assume in 1939 that for any part of mathematics, a formal system could be found for which that part of mathematics would be part of the metatheory.

By the time he wrote [4], in 1963, Curry had changed that definition of mathematics to the "science of formal methods." (This was probably in response to the criticism that under the earlier definition, there could not have been any mathematics before there were formal systems.) This significantly broadens what comes under the definition, and allows for any structure which may be considered "formal." I think that this means that, in today's terms, Curry was a kind of structuralist. For more on this, see [6].

Curry always emphasized that there are two similar notions of truth: 1) truth within some theory (or possibly within some formal system), which is determined by the definition of the theory; and 2) the usefulness of a theory for some purpose. Thus, for example, Curry used constructive (i.e., intuitionistic) logic for the metatheory of combinatory logic and other formal logics, whereas he had no trouble using classical logic as the basis for analysis as used in physics. He did not really believe that there was an absolute truth in mathematics, or in using and reasoning about formal structures.

These structures are clearly created by mathematical activity. But where does this activity come from? How does it originate? And indeed, what does it mean to be *formal*.

One way of looking at this question is to say that it means to look at the *form* and ignore the *content*. For example, counting piles of apples and oranges concentrates not on the kind of fruit but on the numbers of items in the piles. Thus, counting is a formal method.

Counting seems to be almost as old as there have been human beings on Earth. Almost every human language we know about has words for at least the first few numbers. And there is linguistic evidence that numbers have been added over time. In Sanskrit, for example, the word for the number 9, *nava*, has the same root as the word for "new," which can be *nava* or *navaja* or *navajata*. (This is according to http://sanskritdictionary.com.) This is not exactly the same in Latin, which I studied in high school, but the Latin word for the number 9, which is *novem*, does seem closely related to the adjective for "new", which is the adjective *novus, nova, novum*, etc., although it is not one of the forms of the latter.

But it appears that the idea that there were infinitely many counting numbers is not really very old. The first person to write about that was Archimedes, in the third century B.C.E., in his paper "The Sand Reckoner." Archimedes showed this by providing a system for naming natural numbers that clearly can name a number of any magnitude. Perhaps, to use modern terminology and the fact that finite cardinal numbers and finite ordinal numbers are the same, we should call "The Sand Reckoner" a paper on *finite ordinal diagrams*.

It is known that all normal children easily learn the language spoken by the adults around them. This seems to be "hard wired" into us as a species. It also appears that this "hard wiring" can lead to counting. It thus appears that we have "hard wiring" for some mathematical activity. We also seem to be "hard wired" for reflecting and examining everything we do. This suggests that this is where we start doing mathematics; i.e., where we start using formal methods.

I think that Curry's ideas can constitute a kind of bridge between formalists and structuralists.

References

[1] H. B. Curry. Grundlagen der kombinatorischen Logik. *American Journal of Mathematics*, 52:509–536, 789–834, 1930. Inauguraldissertation. Translation into English by J. P. Seldin and F. Kamareddine, *Foundations of Combinatory Logic*, College Publications, 2017.

[2] H. B. Curry. Remarks on the definition and nature of mathematics. *Journal of Unified Science*, 9:164–169, 1939. All copies of this issue were destroyed during World War II, but some copies were distributed at the International Congress for the Unity of Science in Cambridge, MA, in September, 1939. Reprinted with minor corrections in *Dialectica*

8:228–333, 1954. Reprinted again in Paul Benacerraf and Hilary Putnam (eds) *Philosophy of Mathematics: Selected Readings*, pages 152–156, Prentice Hall, Englewood-Cliffs, New Jersey, 1964.

[3] H. B. Curry. *Outlines of a Formalist Philosophy of Mathematics*. North-Holland Publishing Company, Amsterdam, 1951. Mostly written in 1939.

[4] H. B. Curry. *Foundations of Mathematical Logic*. McGraw-Hill Book Company, Inc., New York, San Francisco, Toronto, and London, 1963. Reprinted by Dover, 1977 and 1984.

[5] J. P. Seldin. Arithmetic as a study of formal systems. *Notre Dame Journal of Formal Logic*, 16:449–464, 1975.

[6] J. P. Seldin. Curry's formalism as structuralism. *Logica Universalis*, 5:91–100, 2011.

[7] Stewart Shapiro. *Thinking About Mathematics*. Oxford University Press, Oxford, England, 2000.

Resnik's Structuralism in Light of the History of Mathematics

Ladislav Kvasz
Institute of Philosophy of the Czech Academy of Sciences and Pedagogical Faculty of Charles University in Prague

Abstract

The aim of this paper is to juxtapose with the history of mathematics the basic principles of structuralism presented by Michael Resnik in his *Mathematics as a Science of Patterns*. This juxtaposition shows that Resnik's structuralism has sufficient conceptual resources to offer a plausible interpretation of the historical material, whilst also pointing to some problems within Resnik's theory.

Keywords: Resnik, Structuralism, History of Mathematics, Idealization, Naturalism, Indispensability, Abstract Objects

Structuralism is today – alongside logicism, realism, naturalism and nominalism – one of the main positions in the philosophy of mathematics (see [34]). Geoffrey Hellman, in his overview [12], distinguished four kinds of structuralism: set theoretical structuralism, *sui generis* structuralism, structuralism in category theory and modal structuralism. This paper deals with a variant of *sui generis* structuralism developed by Michael Resnik in a series of articles[25, 26, 27, 28, 29] and summarized in *Mathematics as a Science of Patterns* [30]. Its aim is to consider Resnik's position in terms of the history of mathematics. I have applied a similar approach to the philosophies of mathematics of Ludwig Wittgenstein [14], Gottlob Frege [15], Edmund Husserl [16], and Immanuel Kant [18]. Of course, Resnik – just like Wittgenstein, Frege, Husserl and Kant – is a philosopher; therefore to look in his position for historical inaccuracies would not be particularly illuminating. However, that is not my goal, just as it was not in the case of Wittgenstein, Frege, Husserl, and Kant. Besides the question of historical accuracy, which in philosophical texts is usually marginal, we can confront a philosophical position with the history of mathematics in a more interesting and relevant, i.e. *conceptual* way.

The author expresses his gratitude for the generous support of the project, *Formal Epistemology – the Future Synthesis*, within the framework of the *Praemium Academicum* program of the Czech Academy of Sciences.

When a philosopher outlines his position in the philosophy of mathematics, this is usually motivated by specific historical episodes: in Resnik's case it was the discovery of non-Euclidean geometry and the emergence of abstract algebra. From the history of mathematics perspective it is interesting to ask what these episodes have in common, what makes them important for the particular philosophical position, and whether there are further episodes of the same kind not taken into account. The goal is to reach a *conceptual understanding* of the examples discussed in a philosophical position.

Considering historical episodes which a philosopher did not have in mind may show that the position has *broader relevance*. For example, Husserl's reconstruction of the birth of modern mathematical science in *The Crisis of European Sciences and Transcendental Phenomenology* [13] need not remain restricted to Galileo, but can also be used when analyzing of the theories of Descartes and Newton [19]. Sometimes an author formulates a position for a particular discipline, and a consideration of the history of mathematics can reveal a possible *generalization of the position* for other areas of mathematics. For example, Frege's interpretation of the development of arithmetic in his *Funktion und Begriff* [10] can be generalized to geometry [15]. At other times a juxtaposition with history can reveal some preconception of the author. The removal of this preconception may give the philosophical position a form that is *much more natural*. Wittgenstein's picture theory of meaning, for example, can be separated from the preconception of the existence of a single pictorial form and we can incorporate linguistic pluralism directly into this theory. A pluralistic picture theory of meaning allows us to describe the history of geometry as a development of the pictorial form [14]. Finally, a consideration in terms of historical material can help to *delineate the boundaries* of a position. Thus in the case of Kant's philosophy of geometry we can defend Kant's theory as a valid interpretation of a stage in the development of geometry [18].

When I decided to confront Michael Resnik's postion with historical material, my aim was not to criticize the accuracy of the historical episodes reproduced in his work, but to achieve a conceptual clarification of his position. In mathematics we can distinguish several levels of complexity of language (see [20]). We can analyze a philosophical position from the perspective of each of these levels and check how it represents each particular level. To achieve a more vivid picture of these levels, I will describe how they change. The first level concerns **idealizations**. Development at this level connects Euclid's *Elements* (seen as the paradigm of ancient science) with Newton's *Principia* (seen as the paradigm of modern science). Euclidean science (exemplified by Ptolemy's astronomy) looks for a geometrical order to explain the phenomena, while Newtonian science looks for dynamic laws. The objects of the Euclidean idealization are *atemporal*, *acausal*, and *immaterial* whilst the objects of the Newtonian idealization are *temporal* (time acts as a differential operator in the dynamical equations), *causal* (action is represented by forces) and *material* (it uses material parameters such as density or viscosity). We must not forget that despite their temporality, causality, and materiality the objects of Newtonian physics do not cease to be ideal. *Their time* (an operator in the equation) is not our time (of aging and of death); *their causality* (described by forces acting at a distance) is not our causality (connecting our deeds with their effects); and *their materiality* (having every aspect objective and sharp) is not the materiality of the flesh with its subjectivity and fuzziness.

The second level concerns **instrumental practice**. Development at this level connects Euclid's *Elements* (seen as the paradigm of synthetic geometry that constructs its objects using compass and ruler) with Descartes' *La Géométrie* (seen as the paradigm of analytic geometry that constructs its objects by means of a coordinate system) and Mandelbrot's *The Fractal Geometry of Nature* (seen as a paradigm of iterative geometry, that constructs its objects as a limit of a convergent iterative process). The objects studied by mathematics are not immediately accessible. Therefore mathematicians create representational tools. On the one hand we have *instruments of symbolic representation* such as the decimal positional system in elementary arithmetic, matrices in linear algebra and Leibniz's symbolism in differential and integral calculus. On the other hand, we have *instruments of iconic representation* such as ruler and compass constructions in synthetic geometry, in analytic geometry the construction of curves by means of plotting them in a coordinate system and constructions of figures by means of an iterative process in fractal geometry. Although numbers and triangles as mathematical objects are atemporal, acausal, and immaterial, the digits of a decimal inscription or ruler and compass constructions are pigment marks on paper. They are *temporal* (we can draw and delete them), *causal* (they act on our senses), and *material*. Even though mathematical objects are atemporal, acausal, and immaterial, their representations are temporal, causal, and material.

The third level of linguistic complexity concerns **conceptual frameworks**. Development at this level connects Euclid's *Elements* (seen as an interpretation of the Euclidean geometric universe) with Lobachevski's *On the foundations of geometry* (seen as an interpretation of the non-Euclidean geometric universe). Euclidean geometry interprets straight lines passing to infinity differently from the Bolyai-Lobachevski geometry. At this third level an important developmental stage was marked by Klein's *Erlanger program*, which was one of the first examples of the structuralist approach to mathematics. This approach found its expression in van der Waerden's *Moderne Algebra* and its culmination are the *Éléments de Mathématique* by Nicolas Bourbaki. It is not possible to describe these three levels in detail in this paper. However, the first of them is described in [19], the second in [15] and the third in [14].

In the following I will juxtapose Resnik's position with these three levels of complexity of the language of mathematics, establishing that Resnik's theory contains a plausible interpretation of each, thus rendering his theory an attractive, promising project in the philosophy of mathematics. On the other hand, relating Resnik's theory to the history of mathematics also enables a three–level view of his theory. At each of these levels Resnik's theory has faced particular criticism. I believe that after disentangling the three levels in Resnik's theory it will be possible to address the critical objections. If these objections are expressed in general terms, they seem plausible; yet if we put them into the proper historical context, they can easily be answered.

1 Resnik's structuralism from the idealizations perspective

In Resnik's theory the level of idealizations is present in the form of the indispensability argument and in the interpretation of the boundary separating mathematics from physics. Resnik precisely formulates the indispensability argument by separating its three aspects ([30], p. 45):

"***Indispensability:*** reference to mathematical objects and reliance on mathematical principles is essential to the practice of the natural sciences.

Confirmation holism: observational evidence in favor of a particular scientific theory concerns the theoretical apparatus as a whole, rather than individual hypotheses.

Naturalism: Science is the final arbiter of truth and existence."

On the basis of these points, Resnik formulated **the indispensability argument for mathematical realism**: "*Mathematics is an indispensible component of the natural sciences; due to holism, evidence in favor of science is also evidence in favor of mathematical objects and mathematical principles that the theory presupposes; thus, as the consequence of naturalism, mathematics is true and the existence of mathematical objects is as well justified as the existence of other entities posited by science.*" Resnik calls his argument a Holistic-Naturalistic (H-N) argument about indispensability. I shall now seek to respond to a series of objections.

1.1 The question of the causal inertness of abstract objects

In his review, Mark Balaguer cites, as the main argument against Resnik's position, the causal inertness of abstract objects: *"It seems to me that even if our mathematical theories are indispensable to the descriptions and inferences of empirical science, this gives us no reason whatsoever to believe that these mathematical theories are literally true or that there are any such things as abstract mathematical objects. The reason, in a nutshell, is that abstract objects are supposed to be causally inert."* ([2], p. 113). Balaguer is right. His argument shows that the interpretation of mathematical objects as *causally inert, abstract things* is mistaken. This interpretation was created in the times of Plato and Aristotle, that is, at a time which considered the only ideal objects to be the atemporal objects of arithmetic and synthetic geometry. It was that context which gave birth to the view that *the ideal objects of mathematics are abstract and causally inert*. Aristotle's theory of abstraction is, at least in the case of mathematics, *a subjectivist interpretation of the process of idealizations*. However, if we consider a real marble, such as the one used by Galileo in his experiments with an inclined plane, and we start abstracting, it is not clear where we end up. A geometer will abstract not only from color, taste and smell, but also from elasticity, weight, heat capacity and electrical resistance, and will retain only shape and size; a Newtonian physicist will retain also weight, density, and elasticity, but abstract from the electrical resistance, heat capacity and other qualities of the ball. Yet how do they know where to stop? The only plausible explanation is that in abstraction we retain what is dictated *by the language* of geometry or mechanics, in order to create a geometrical or a mechanical representation of the particular object.

Thus we are dealing not with a *psychological process of abstraction*, as the introduction of ideal objects was explained by Aristotle, but with an *epistemological process of linguistic reduction*. When we create a linguistic representation of reality, we ignore all those aspects that the formal language (of geometry or mechanics) cannot represent. At the time of Aristotle, there was practically only one formal language (which a century later was codified by Euclid in his *Elements*). Therefore, Aristotle could not recognize the linguistic nature of this reduction, so he interpreted it psychologically as abstraction. If we want to be faithful to *Frege's rejection of psychologism in the philosophy of mathematics*, we must also reject Aristotle's theory of abstraction. **Objects of mathematics are not abstract but ideal objects.** Thus the question of the causal inertness of abstract objects leaves the philosophy of mathematics untouched. *Idealization is a process of linguistic reduction*, in which the syntax of a fragment of language is reduced to a set of formal rules. Ideal objects are objects to which the expressions of the language refer after this reduction. Therefore, ideal objects are not necessarily causally inert - at least, not since

Newton, who created a formal language for representing causal interactions between bodies as actions of forces. This formal language, in contrast to that of geometry, retains the ability of ordinary language to refer to causal relations.

The reason why the ideal objects of mathematics are considered by most philosophers of mathematics to be abstract and causally inert is probably that at the time when Plato and Aristotle created their interpretation of the ideal objects of mathematics, the only known kind of idealization was the one codified by Euclid. The ideal objects introduced by the Euclidean idealization are the causally inert objects of Euclidean geometry, such as triangles and circles. This idealization was the only one known even up to the beginning of the 17th century, and so according to Galileo the book of nature is written in terms of triangles and circles. Galileo still understood idealization in a Euclidean spirit - but was one of the last prominent scientists to see it that way.

It is easy to realize that the systems described by Galileo – free fall, the pendulum, projectile motion – are systems consisting of a single isolated body. We see that *Galileo was unable to describe interaction.* In his physics Galileo reached the limits of Euclidean idealization. Shortly after Galileo came Descartes and Newton, who introduced causally acting ideal objects. Descartes' extended body is a mathematical ideality, that is, an object introduced by the same linguistic reduction as the perfect geometric sphere discussed above; but unlike the ideal objects of geometry Descartes' extended bodies are mobile and have inertia (the tendency to remain at rest or in uniform rectilinear motion) and the ability to interact with other bodies, that is, to causally affect other extended bodies. Newtonian bodies are able even to act at a distance (for details see [19]).

The Newtonian idealization introduced a completely new kind of ideal object, that is, ideal objects capable of causal interaction: thus the causal inertness of abstract objects is a myth. Since at least the time of Newton it has been a fallacy that abstract objects are causally inert. In physics we have plenty of abstract objects which are *not* causally inert. Newtonian mass points are abstract objects, just like Euclidean triangles but despite being abstract objects, mass points act on each other by means of forces acting at a distance: thus to try to disqualify all abstract objects because of their alleged causal inertness is a mistake. There are abstract objects, such as material points, incompressible fluids, or electromagnetic fields, which are not causally inert, so Balaguer has an excessively simplistic idea of abstraction. He must draw a boundary between the Euclidean and the Newtonian abstraction. As long as this is not done, his argument misses its target.

1.2 The boundary between the abstract and the concrete

Balaguer understands the role of mathematics in the Galilean sense: mathematical objects are artifacts forming a *language used in the description of nature*, but they do not enter into nature's constitution. However, when Descartes and Newton wanted to describe interaction they introduced into the universe of ideal objects *time*, *motion*, and *interaction*. Newtonian bodies subjected to forces acting at a distance are ideal objects. However, they are ideal objects of a new kind that Newton introduced in order to mathematically represent causal interactions among bodies. Balaguer ignores this development and writes: "*insofar as the realistic view takes abstract objects to be causally inert, it predicts that if there were no mathematical objects, the physical world would be as it is right now. Thus, it seems to me that whatever empirical data we are receiving right now, we cannot take these data to support the existence of abstract objects, for even if there were no such things as abstract objects, we would still be receiving these same data.*" ([2], p. 113).

This is, of course, a paraphrase of Galileo's famous argument about colors, tastes and odors from *The Assayer*: "*Hence I think that tastes, odors, colors, and so on are no more than mere names so far as the object in which we place them is concerned, and that they reside only in the consciousness. Hence if the living creature were removed, all these qualities would be wiped away and annihilated.*" ([11], p. 274). Galileo used this argument to show that colors, tastes, and odors are secondary qualities. When later Newton, using a prism, made colors measurable, he showed that the differences between colors led to different physical events and thus were real. Photosynthesis, the photoelectric effect, and many other processes that are sensitive to color differences show that colors are properties that have causal effects (just like mass, shape, or size), despite the fact that the Galileo was able to abstract from them. To abstract from colors, that is to destroy all differences between the various kinds of light, would have disastrous consequences: photosynthesis would stop and Galileo would as a result of his reckless act of abstraction die form a shortage of oxygen.

Thought experiments - in which we abstract from certain features of reality and imagine that the world continues to function as before - are inconclusive. In my opinion, Balaguer's argument is equally inconclusive. If everything that Balaguer calls an abstract object were to disappear from the world, forces acting at a distance would also disappear and we would soon freeze, because the Earth would depart from its orbit of the Sun. The objection that the disappearance of Newtonian forces would have no effect, because Newtonian physics is not true, is not valid. All physical theories use idealizations. Therefore Balaguer's argument is sustainable only if we understand it as an argument not against mathematical but against scientific realism.

Therefore, Resnik's strategy that Balaguer ironically called *"blurriism about the abstract- concrete distinction"* ([2], p. 109) is correct. Mathematics is not separated from physics by any sharp boundary, and Newton, Euler, Lagrange and Poincaré successfully bridged these two aspects of our understanding. A theory of Newtonian idealization (see [19]) enables Resnik's theory to be furnished with details and thus increase its plausibility. However, even without this further development, Resnik's theory is correct because contemporary physics uses ideal objects that act causally. For Euclid, a triangle can neither move uniformly in a straight line nor collide with a circle. The rules of Euclidean geometry do not permit this. A circle can intersect the sides of a triangle. In contrast, Newtonian particles can move and interact. Yet if we accept that there are fields (i.e., if we take a realistic approach to the theoretical entities of physics – which is today standard – see [24]), we should also take a realistic approach to objects of Euclidean idealization.

1.3 Space and time as criteria of concreteness

A common objection to a realistic interpretation of mathematics is that the real objects exist in space and time, whilst the objects of mathematics are abstract and thus outside of space and time. This objection was formulated in 1973 by Paul Benacerraf in his paper, *Mathematical Truth*, and Balaguer used it against Resnik's structuralist claim that *"abstract objects never exist in space-time and could not exist in space-time."* ([2], p. 114). Resnik accepted this argument and admitted that mathematical objects were acausal and outside space and time ([30], p. 82). Nevertheless, I would like to stress that this is a strange argument. Space and time are indubitably mathematical constructs. As put by Carl Friedrich von Weizsäcker: *"A glance at the history of philosophy and of physics shows that space that was one hundred years after Newton considered an a priori aspect of our ability to know, was one hundred years before Newton known by nobody."* ([37], p. 203). Space is a mathematical object (for Newton with a theological touch of the *Sensorium Dei*). Space is three-dimensional, homogeneous, isotropic, infinite, and continuous, which are mathematical characteristics. Therefore, if someone claims that the objects of the real world exist in space and time, he espouses a form of mathematical realism.

Newton "inserted" the real world into *true, absolute, mathematical space*; he replaced real objects by bodies with sharp, unequivocal, measurable characteristics represented by mathematical quantities and endowed them with *forces acting at a distance*. It was action at a distance that forced Newton to "insert" the real world into empty space. Nevertheless, action at a distance contradicts common sense. Therefore to declare the basis of action at a distance – the empty space – a criterion of reality, needs substantiation. Space and time are mathematical objects, so this may be an internal mathematical dispute. One mathematical object (space-time) is declared an arbiter of reality of other mathematical objects. Some mathematical objects – numbers, triangles, and polynomials – are denied the status of reality because they exist outside this privileged mathematical object while other equally mathematical objects – forces, fields, and quanta – are declared real, because they exist inside this privileged mathematical object. It is far from clear why space-time is so privileged.

Existence in space and time can be understood also metaphorically. The claim that real objects exist in space can be understood in the sense that these objects can be localized relative to other objects. Consequently, some other mathematical objects, such as numbers and sets, are denied real existence because they cannot be so localized. That sounds at first glance plausible, but what exactly is meant by localizing, what instruments are allowed, and what accuracy is required? One of the properties of quantum-mechanical objects is their non-locality. Our criterion would therefore rule out the existence of any quantum-mechanical object. Macroscopic objects, which the advocate of the criterion of localization probably considers as really existing, are composed of electrons and protons, which he would have to declare non-existent. In addition, some objects of *Euclid's* geometry meet Balaguer's criteria. Geometrical objects are located in space (e.g. a sphere) and since *Euclid's* first three postulates have a constructive character, they are executed over a passage of time. (I assume it is not possible to join two points with a straight line without the elapse of some time). Therefore some objects of geometry are objects which exist in space and time, and thus real.

2 Resnik's structuralism from the perspective of re-codings[1]

In the fifth chapter of his book, Resnik passes to the question of how we can acquire knowledge of mathematical objects. He responds Paul Benacerraf's *Mathematical Truth* [4], which argues in favor of the thesis that mathematical objects do not have causal effects on our senses and therefore we cannot acquire knowledge of them. Benacerraf's arguments are, according to Resnik, flawed. Nevertheless, they do show that realists need to develop an epistemology of mathematics. Resnik's approach to this task is based on challenging the sharp difference between mathematics and the natural sciences. He stresses the holistic nature of evidence in both science and mathematics. In the ninth chapter Resnik presents a quasi-historical interpretation

---

[1] Re-codings is the name given in [17] to complete changes of instrumental practice, such as the replacement of ruler and compass constructions by coordinate diagrams of analytic geometry.

of two systems, those of arithmetic and of geometry ([30], pp. 177-182). He argues that the *language* created by means of these systems *refers to mathematical objects*.

In the theory of re-codings (see [17], pp. 11-84) in addition to the two systems described by Resnik, six more are added. If we plot these eight representational tools according to the historical order of their introduction, several interesting phenomena appear. These phenomena escaped Resnik's attention. The first is a phenomenon which we shall call the **principle of bipolarity** – the language of mathematics has two poles, one symbolic and one iconic. In the development of mathematics these two poles regularly alternate. After an *instrument of symbolic representation* (such as the decimal positional system in arithmetic, the notation system of polynomials in algebra, or Leibniz's notational system of the differential and integral calculus), next there is introduced an *instrument of iconic representation* (such as ruler and compass constructions in synthetic geometry and constructions by means of a coordinate system in analytic geometry), and vice versa.

Resnik also noted the existence of representational tools: structuralism *"enters my explanation of how our handling the specific numerals and diagrams can shed light on the abstract realm of mathematical objects. Here the basic idea is that these particular tools represent the abstract structures, which we study."* ([30], p. 7.) The problem is that Resnik understands the *relation of representation* not in the linguistic sense as a relation, constituted by the rules of a language game, but ontologically, as instantiation, a point also made by Janet Folina ([9], p. 471).

It is important to understand that the same mathematical content can be expressed using different instruments. Addition, subtraction, multiplication, and division can be expressed by *an instrument of symbolic representation* (in arithmetic), i.e. a manipulation with number symbols. However, the same result can be obtained also using *an instrument of iconic representation* (in synthetic geometry) i.e. a manipulation with numbers represented by line segments. It is interesting that the product of numbers a and b can be represented in two ways: firstly, as in Euclid, as the area of the rectangle with sides a and b, and secondly, as in Descartes, as a line segment the length of which is a times b. This phenomenon can be called **mirroring** – contents that were for the first time expressed by means of a particular instrument are mirrored in the universes of the later instruments, and often in several different ways. In *What Numbers Could Not Be* [3], Benacerraf drew attention to the phenomenon of mirroring in the case of arithmetic and set theory. Nevertheless, set theory is by no means exceptional: mirroring is typical for all representational tools.

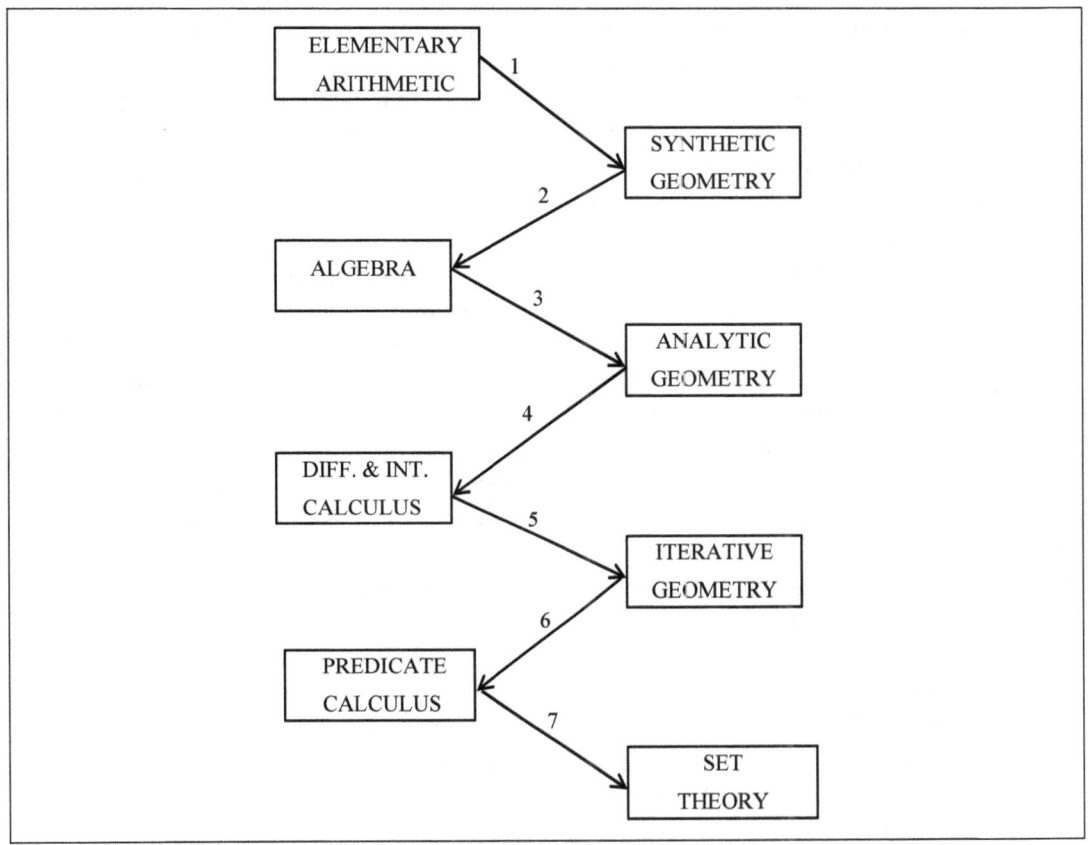

Figure 1

2.1 New mathematical objects and appropriate instrumental practice

Bipolarity reveals a further phenomenon that is important for an understanding of the epistemic aspect of mathematics: *first objects*, for the representation of which a new representational tool will be created, emerge *in the universe of the previous representational tool of the same kind*. For example, **the first curves** – the *conchoid*, the *cissoid*, the *quadratrix*, and the *spiral* – emerged *in the universe of synthetic geometry* i.e. the universe of objects constructed by means of the compass and the ruler. For their representation many years later the representational tool of **analytic geometry** was created. Thus we cannot claim that in synthetic geometry there were no curves – there was *one* cubic curve, *one* biquadratic curve and *two* transcendent curves. Of course, in this respect synthetic geometry cannot be compared with analytic geometry, which distinguishes *78 kinds of cubic curves* (72 of them were found by Newton) and many more curves of higher degrees, of which the ancient mathematicians had no idea (see [5]). Thus synthetic geometry cannot be compared with analytic geometry in its expressive power, although that is not at issue now. The fact that I want to emphasize is that *the first objects of the new kind* (i.e. the first planar curves), for the study of which Descartes *introduced a new representational tool*, emerged *within the universe of the previous representational tool* of the same kind (i.e. the previous tool of iconic representation).[2] I suggest calling this phenomenon **the principle of the pre-existence of objects** *(of a certain representational tool in the universe of the previous tool of the same kind)*.

This principle is crucial for understanding the construction of new representational tools. It shows that mathematicians have some (even if, in hindsight, fairly limited) knowledge of the mathematical objects, before they introduce the new representational tool designed for their study. By a **representational tool** I mean a *system of physical objects* – usually ink marks on paper. The manipulation of these objects is subordinated to the *rules of a language game*. However - and at this point I disagree with Resnik - these rules are *conventional*. They are designed so that the functioning of the tool *copies the properties of the mathematical objects* which the tool is designed to represent (this being possible as the mathematicians already have knowledge of these objects, at least to a limited extent, due to the pre-existence of the particular objects). For example, Fermat discovered the rules for the derivation of quadratic and cubic functions some fifty years before the formation of the differential and integral calculus. He used the instrument of the symbolic representation of algebra. When Leibniz created the representational tool of the differential and integral calculus, he created it (that is he introduced a *set of conventional rules* for handling chains of ink marks) so that these rules *reproduced* Fermat's rules for the derivation of polynomial functions. The representational tool developed by Leibniz brought a dramatic increase in the expressive power of language and brought substantial progress in the study of functions. This is so because this tool was designed so that functions and, first of all, operations with them (i.e. differentiation and integration) could be easily handled; and Leibniz could create it because to a limited extent he already knew the rules for differentiation and integration.

Representational tools can be compared with measuring instruments in physics. Before it was possible to create the first thermometer, physicists had to know something about heat and temperature, even though their understanding of these phe-

---

[2] Just like the first functions, that is the first objects for the study of which the differential and integral calculus was created, were introduced in the universe of algebra, that is the previous tool of symbolic representation, in the form of polynomial functions and rational functions. This phenomenon seems quite general.

nomena was limited. They created the thermometer in order to be able to measure temperature more precisely. Similarly Descartes developed his new tool for representing curves on the basis of his knowledge of the curves of ancient geometry: he created the new tool in order to handle curves more easily than the Ancients and to study them more efficiently. The *principle of the pre-existence of objects* makes it possible to justify the claim that the new instrument of analytic geometry is a tool that serves the study of curves or functions. Curves and functions exist independently of the representational tools by means of which we study them. Of course accuracy, fineness, and richness of detail increase dramatically after the introduction of the new tool, but the objects themselves are independent of the particular instrument. That the derivation of cosine gives minus sine is a fact that does not depend upon differential calculus. By means of Leibniz's calculus we can prove it in one line, but it can also be derived by means of geometry. Mathematics studies an independent reality by means of instruments of symbolic and iconic representation, just as is claimed by Resnik.

This is important because it clarifies a point upon which foundered Penelope Maddy's project of a realistic interpretation of mathematics. Maddy tried to anchor the language of set theory (which is, according to the theory of re-codings, the eighth representational tool) in ordinary experience. According *the principle of the pre-existence of objects*, this is wrong because the first objects of set theory were born in the universe of the previous representational tool of the same kind, which is iterative geometry. (In the theory of functions of a real variable, functions are often constructed as the limits of some iterative process). Bolzano, Dedekind, and Cantor came to their versions of set theory through the analysis of problems of the theory of functions of a real variable. *In the universe of the theory of real functions it is possible to motivate the introduction of most of the postulates of set theory.* Therefore, I think Maddy took a false step when she declared that in a refrigerator we have *a set of eggs*. The critics of her theory attacked this point with great vehemence. However, whether or not we have in our refrigerator a set of eggs has no influence on the realistic status of the objects of set theory. *The universe of set theory* should be anchored[3] in the universe of the theory of functions of a real

---

[3]With anchoring I mean the motivation of the introduction of the syntactic rules of the new representational tool by knowledge of the objects for the study of which this new tool is constructed. (Motivation of the rules is of course something much weaker than their justification, but on the basis of limited knowledge nothing more is really possible.) This knowledge is obtained by means of the previous tool of the same kind (i.e. symbolic or iconic). Therefore I say that the rules of the new instrument are anchored in the knowledge gained by means of the previous instrument. Thus the differential calculus of Leibniz was anchored in the knowledge of derivations that Fermat gained by means of algebra, i.e. by means of the preceding instrument of symbolic representation (i.e. the instrument of the same kind as the calculus).

variable; *the universe of the theory of functions of a real variable* should be anchored in the universe of analytic geometry; *the universe of analytic geometry* has to be anchored in the universe of synthetic geometry and *only the universe of synthetic geometry* must be anchored in the world of our ordinary experience.

Maddy made a mistake when she claimed that sets can be found *immediately* in the world around us. They cannot. Yet in spite of that, by means of appropriate *mediation* it is possible to pass from the universe of set theory to the world of our ordinary experience. Of course, it is possible to *describe* the world of our ordinary experience by means of sets but sets are *not anchored* in this world; therefore particular set-theoretical distinctions (e.g. between the empty set and the singleton containing the empty set) cannot be motivated by ordinary experience, but only by the phenomena of the universe of real functions.[4]

The *anchoring* of the universe of a certain representational tool in the universe of the previous tool of the same kind and the *derivation* of the conventions that constitute the new representational tool from the experience with mathematical objects belonging to the previous universe allows us to explain how mathematicians obtain knowledge of mathematical reality. Lines or symbols drawn on paper are **physical objects that can be seen.** They are ink marks that can be placed on paper and then examined. *Conventions to which we subordinate the practice* of geometrical constructions and symbolic manipulations are derived from the properties of mathematical objects, for the study of which we created the new tool of symbolic or iconic representations. For example, when we look at a tangent to a curve, despite the fact that on paper the area of contact between them appears to have shape and size, in our proofs we follow the conventional rule (which we base on the properties of contact between two mathematical curves) that the contact between two curves is an indivisible point without shape and size. The shape that we see is interpreted as a manifestation of the imperfection with which our representational tools represent the corresponding mathematical situations.

---

[4] Thus if the Ancient Greeks had possessed refrigerators containing cartons with eggs, they would still not have needed to develop set theory. Elementary arithmetic would have been fairly sufficient. However, if they had become interested in the question of the convergence of Fourier series, it is likely that sooner or later they would have developed something similar to set theory. Of course we can describe cartons with eggs using set theory, because we can describe almost everything by means of this theory. Set theory is the mathematical theory with the highest expressive power. However, we cannot motivate the concepts and principles of set theory by means of cartons of eggs.

2.2 The anchoring of elementary arithmetic and synthetic geometry

We consider the *geometric figures* constructed by means of an instrument of iconic representation as well as the *symbolic formulas* constructed by means of an instrument of symbolic representation to be elements that belong to the tools by means of which we represent mathematical reality, and not to the mathematical reality itself. These elements *do not form the subject matter of mathematics*, but they make it possible for us to study this subject matter – curves, functions, or sets. It is thus possible to defend the realistic status of the universe of mathematical objects studied by the various representational tools, but only from the third tool onwards (i.e. from the objects such as polynomials or matrices, studied by the instrument of symbolic representation used in algebra). The *first instrument of symbolic representation*, that is, the numerical system of elementary arithmetic, and the *first instrument of iconic representation*, that is, the geometrical system of ruler and compass constructions, are exceptions. For them there is no previous representational tool of the same kind that we could use to anchor them.

If we want to create a plausible epistemology of mathematics, we also have to interpret the syntactic rules of elementary arithmetic and the construction rules of constructions of elementary geometry as *conventions* introduced on the basis of a knowledge of mathematical reality. Only then can a realistic interpretation of the epistemological aspects of mathematics begin. Resnik offers "imaginary histories" and Maddy a theory that relies on cognitive psychology research. Resnik's histories appear to be questionable, but Maddy's interpretation can be accepted. The criticism provoked by her theory was directed against the idea that what we recognize (for instance in our refrigerator) is a set. However, if we interpret the propositions regarding small collections of objects not as facts about sets (i.e. objects of set theory), but as facts about numbers (i.e. objects of elementary arithmetic), it is possible to accept Maddy's anchoring of this theory in our cognitive capacities. The point is that in everyday life we have enough experience with small numbers and elementary geometric shapes to base the first two representational tools on this experience.

If we consider figures drawn by means of ruler and compass not as the subject matter of geometry, but as elements of the representational tool, we need not pretend that two straight lines drawn on paper intersect at a single point. Since they are elements of the instrumental practice created by us, we can introduce conventions to interpret the expressions constructed by means of these tools. Thus if on paper we see that the intersection of two straight lines looks like a diamond, we need not pretend that in the case of a perfect ruler it would be a point. It is sufficient that **we will use** the intersection of the lines drawn on paper **as a representation of**

the point at which the geometrical lines that we represent by means of the ruler intersect.

2.3 Axiomatic theories as a source of knowledge

Resnik's explanation of how we acquire knowledge in mathematics is based on the idea that axiomatic theories provide us with knowledge of mathematical structures because they provide their implicit definitions. Even though this explanation has been praised (e. g. [2], p. 120), from the history of mathematics perspective it contains a tension. The axiomatization of a theory comes usually long after its basic theorems have been proven so we must have knowledge prior to the axioms. One possible way to overcome this tension is to see the axioms not in the *logical sense* as propositions functioning as premises in proofs, but in the *instrumental sense* as principles making explicit the rules of the instrumental practice. The instruments themselves open the epistemic access to the objects of the mathematical universe and the axioms simply codify the conventions that constitute them.

The instruments of symbolic and iconic representation are real objects – in the case of synthetic geometry the instrument consists of a ruler made of wood or plastic, a compass typically made of iron, and of course a pencil and paper. When we use them, we place traces of graphite on paper. These traces are physical things, so we can causally interact with them - we can draw them, see them, and erase them. Nevertheless, we *subordinate our instrumental practice and the interpretation of its results to the principles of Euclidean geometry*. These principles are first implicit, embodied in the instrumental practice, but later they take the explicit form of axioms (or postulates). Thus when we erect on a given straight line a perpendicular, we maintain (in accordance with Euclid's fourth postulate) that the four right angles are identical, even if strictly speaking they are not and cannot be, because at the microscopic level the traces of graphite on paper are most irregular. Similarly, when two straight lines intersect, we say that they intersect at a point, even though we see that their intersection has the shape of a diamond. In other words, we use the compass, the ruler, and the pencil as *elements of a language game* which are subjected to certain rules - and the rules of the language game are chosen such that they agree with the *principles of Euclidean geometry*.[5]

---

[5] At first glance this behavior may seem incomprehensible – mathematicians, contrary to what they actually see, speak about things which they do not see and which, strictly speaking, cannot be seen. We must not forget, however, that the axiomatic method was introduced in response to the discovery of incommensurability (and of Zeno's paradoxes) and in a situation where intuitively constructed systems have proven to be inconsistent, it is a natural reaction to subordinate intuition to publicly controlled rules, which are called axioms and postulates.

Thanks to the subordination of the rules of the language game to the principles of Euclidean geometry, the instrumental practice of the constructions with ruler and compass allows us to study the properties of geometric objects. The objects that we study in geometry are ideal points, lines, and circles. Of course, what are really present on paper are not ideal points, but pieces of graphite. Instead of an infinite straight line we have a finite number of chunks of graphite. The important thing is, nevertheless, that *we use these pieces of the graphite within the instrumental practice of the language game of synthetic geometry so as if they were ideal.* Geometry studies not the traces of graphite on paper, but by means of these traces it represents the ideal geometrical objects that form the subject-matter of her interest. It can do this because Euclid's postulates enable us to check each step of a construction or proof and identify any deviations from the representation of an ideal geometrical object. If the deviation is disturbing, the representation can be remade - the scales can be changed, the relative positions of the objects can be altered, and the viewpoint can be shifted.

Thus using instruments of symbolic and iconic representation we study ideal mathematical objects.[6] The fact that the intersection of two transversal lines is a point is simply a rule of the language game to which we must subordinate the contingent empirical fact that the intersection appears as a diamond. Anyone who understands the rules of the language game knows that it is a point, no matter how it looks. Because the language game has a normative dimension, it opens the door to a knowledge of mathematical objects. This interpretation is close to Resnik's. The difference is that instead of interpreting the axioms as implicit definitions of structures of "abstract" objects, I see them (or some of them) as rules constituting a language game and it is the instrumental practice constituted by this language game that gives us access to mathematical knowledge.

2.4 Representational instruments and intuition

Mathematical objects are not actually present in the instrumental practice (they are not present in the straight lines and circles drawn on paper), *but they constitute the normative aspect of this practice.* We *subordinate* the instrumental practice to norms derived from the properties of ideal objects that appeared in the universe of the previous instrumental practice. In all constructions and arguments *we use*

---

[6]We have to keep in mind that in ancient Greece, mathematicians had at their disposal no means other than the tools of elementary geometry and arithmetic. The works of Archimedes, Apollonius, and Euclid show that by these means they proved rather complicated theorems. It is not possible to open the question of the rigor of their proofs here, but there is a sufficient literature devoted to that issue (see [23]).

points as objects having no parts and *we use* straight lines in accordance with the rules of geometry. A straight line drawn on paper is not only *a physical object*, but it is also *a sign*, i.e. an element of a language game. That means that for a figure drawn on paper we can use only properties that are sanctioned by the postulates of geometry. If we follow these postulates, interpreted as rules of the language game, we can gain knowledge about ideal mathematical objects. Gaining mathematical knowledge is possible because we can causally interact with the instruments of symbolic and iconic representation; and if we subordinate these instruments to the rules of mathematics, we can obtain from our causal interactions with these instruments knowledge about mathematical objects, their properties, and relations. For example, we can use numerals written on paper to verify arithmetical facts. Therefore it is irrelevant whether numbers are causally inert or not. It is sufficient to *subordinate* our manipulations with *numerals to the laws of arithmetic*; and we can study numbers despite the fact that with numbers we cannot causally interact.

The understanding of this situation is hampered by the fact that *after some time the rules of the instrumental practice are interiorized*. This internalization allows us to "guess" the outcome of a construction before the construction was made. The interiorization creates *a kind of intuition* associated with the instrumental practice. So *arithmetical intuition* arises from the internalisation of the rules of manipulation of numerals; *geometrical intuition* arises from the internalisation of the rules of ruler and compass constructions; *structural intuition* arises from the internalisation of the rules for manipulations of algebraic expressions. Philosophers like Plato, Descartes, and Kant, and mathematicians like Poincaré, Brouwer, and Gödel, tend to separate the mathematical intuition from its instrumental roots and consider it as the source of epistemological access to mathematical reality.

2.5 A difference between abstraction and idealization from an epistemological viewpoint

In order to question the possibility of studying mathematical reality by means of concrete objects, Balaguer wrote: "*Human beings could not learn anything about any abstract pattern by perceiving a system of concrete objects unless they knew in advance that the given abstract system stood in some particular relation to the given system concrete objects. But how could human beings ever know this? Since they have no epistemic access to abstract patterns, it seems that they could not.*" ([2], p. 122). If we compare this passage with the above explanation of how learning about ideal mathematical objects is possible, we see that the problem with Balaguer's exposition is that he considers the particular concrete objects by means of which we acquire mathematical knowledge not as a representational tool but as *instantiations* of the "abstract patterns". Thus in Balaguer's exposition, human beings have to know that the given concrete objects are in a particular relation to the given "abstract system", and it is not clear how could they acquire this knowledge. In the above explanation the concrete objects form a representational tool, and thus they stay in a particular relation to the ideal objects simply because the mathematicians have chosen the rules that constitute this instrument in such a way: this is possible because due to the *principle of the pre-existence of objects*, at least some of the objects for the study of which the representational tool will serve are accessible by means of the previous instrumental practice. These pre-existing objects make it possible to "calibrate" the new instrument (i.e. to select its conventional rules). However, that has already been said.

I have returned to this quote because it sheds light on the relation between abstraction and idealization. When we speak about abstract objects, we put the burden of the epistemic access on the shoulders of the particular subject – she must obtain epistemic access to abstract objects which she, because of the causal inertness of abstract objects, can never have. However, if we *replace the psychological theory of abstraction* understood as some kind of mental activity of a subject (in the framework of which the epistemic access to mathematical objects has to be intuition or recollection) *by a linguistic theory of idealization*, we have a theory in which our epistemic access is public. It is constituted by a set of publicly available conventions to which we subordinate the linguistic practice constituted by our representational tools. This practice concerns a linguistic community (and not an individual subject) that is anchored in a shared material environment (and not in the subject's mental world). The members of the linguistic community choose several objects of this material environment to form an instrument of symbolic or iconic representation. They physically manipulate elements of this representational tool and subordinate the rules of these manipulations to shared, publicly accessible conventions constituting a language game. These conventions are chosen in such a way that they "stand in some particular relation to the given abstract system" as Balaguer requires.

One such kind of instrumental practice is counting; another consists of geometric constructions using a ruler and compass. Both are anchored in our material environment, the first in some system of countable (i.e. sufficiently similar) objects such as beans, stones, or lines on paper or in sand, or later in various systems of number symbols (understood in the physical sense as inscriptions); the second in a system of reproducible (i.e. sufficiently regular) objects drawn in sand or on paper. In both cases the instrumental practices are related to some real situations such as accounting, surveying, or architecture that help to test the results, and to check the conventions to which we subordinate the linguistic practice of counting or geometrical constructions. After a shorter or longer period in both forms of instrumental practice, idealization occurs. *I understand idealization as a linguistic reduction*, i.e. a total subordination of instrumental practice to linguistic rules. In the case of arithmetic the idealization means that the system of numbers is opened towards infinity and thus we proceed as if we could count quantities that are far beyond our physical or mental capabilities. One manifestation of the awareness of idealization is Archimedes' work, *The Sand Reckoner,* but we have to remember that the idealization itself must have occurred much earlier. Similarly, in the case of geometry, one aspect of idealization can be seen in Euclid's second postulate according to which any finite straight line can be prolonged. In reality it often cannot: the customs officers at some border might stop us doing so. However, in the language game of geometrical constructions we take literally the rules that constitute this game, ignoring such contingent non-geometrical circumstances as customs officers, the size of the universe and the length of life, which in reality cut down our ability to perform some constructions. We will thus, after the particular idealization, act within our language game, i.e. make the manipulations with our representational instruments in such a way, as if the reality represented by means of these instruments were ideal. This is a crucial point. The objects used in the instrumental practice do not change. We need not pretend that the straight line drawn on the paper is infinitely long, or that we would be able to draw, in idealized circumstances, some infinitely long straight line. We know that the lines used in the instrumental practice are finite and we will never be able to draw infinitely long lines. *What is idealized is not the instruments used in the constructions but the reality to which they refer.*

Our linguistic practice of mathematics refers to ideal objects. Thus *ideal objects* are objects embedded in a publicly-shared language game, objects to which the symbols of the game refer. The abstract objects enter the scene after people reach sufficient cognitive adaptation to this linguistic practice. *Abstract objects* are (mental) representations that we form on the basis of our experience with the particular language game, so for instance we can get used to the practice of counting to such an extent that we can predict the results of counting with small numbers before the actual counting with the material instrument is performed. Similarly, we learn to mentally see the simplest geometrical shapes. Thus we gradually mentally separate numbers and shapes from the instrumental practice in which they are embedded and create abstract ideas of numbers and shapes. And not only that! We acquire an impression that these abstract objects actually form the realm of mathematics – and Benacerraf's dilemma is born. However, if we realize that the abstract objects originate in a particular instrumental practice, all mystery disappears. *The abstract objects are not the subject matter of mathematics.* They are byproducts of its instrumental practice. When we study mathematical reality, we do not investigate abstracted objects, and thus their causal inertness is irrelevant. *The subject matter of mathematics is formed by ideal and not by abstract objects. We study these ideal objects in the framework of a publicly controlled instrumental practice and not by subjective intuition.*[7]

3 Resnik's structuralism from the perspective of relativizations[8]

The theory of relativizations is described in *Patterns of Change*. According to this, the introduction of new objects occurs on the basis a *form of language* that determines the semantic rules for their interpretation. An example of relativization in geometry was the introduction of space (when geometers started to consider empty space as consisting of points), or the introduction of infinity (when they started to consider straight lines of geometry as infinitely extended). Relativizations in algebra were, for example, the introduction of complex numbers and the reification of algebraic structures. I suggest distinguishing eight *forms of language*. There is no space for their description here, but I will give a brief overview (see Figure 2):

1. **The perspectivist form**: language represents a picture of reality existing *independently* of language; the *picture* of reality is constructed from a *fixed viewpoint* that is not included in the picture.

2. **The projective form**: the viewpoint is incorporated into the picture; it is possible to *relate pictures*, constructed from different viewpoints, to each other.

3. **The coordinative form**: language is able to coordinate many elements created from different viewpoints and to unite them by a principle of coordination to form *a map* of reality.[9]

---

[7] To some extent this resembles Hilbert's approach to ideal objects. The difference is, of course, that Hilbert developed his approach as a mathematical method the aim of which was to prove the absolute consistency of some theories. In my interpretation it is a philosophical account of how mathematics works. Thus even if Hilbert's hopes were destroyed by Gödel, the above account could still be plausible. Even if it cannot be used for proving absolute consistency, it still can be correct.

[8] Relativizations is the name given in [17] to radical changes of the conceptual framework of a mathematical theory, in the course of which the instrumental practice undergoes only small changes, which do not disturb its continuity.

[9] This form of language is used in cartography. There if we construct a map of Africa, the shape we see is not the reproduction of something that somebody could see from a particular viewpoint. Before space flight it was not possible to see the shape of Africa. The shape of Africa on maps cannot be interpreted as a picture in the sense of the perspectivist form – i.e. a reproduction of an image that someone has seen. The shape of Africa is a map, i.e. an image created by a coordination of a great number of pieces of information coming from different people.

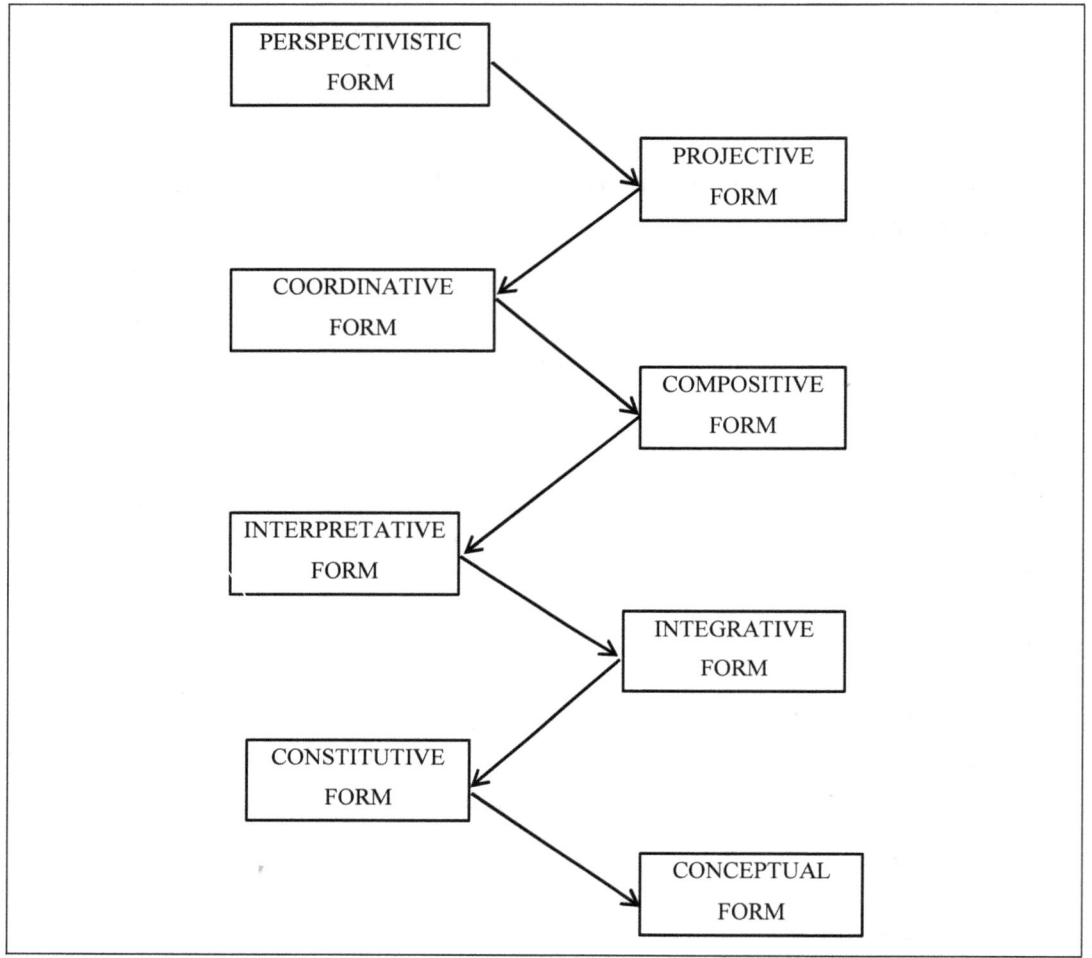

Figure 2

4. **The compositive form**: the viewpoints, the information gathered from which was coordinated in the previous form, had identical epistemic structure. The compositive form allows the unification of images created by different kinds of epistemic subjects. These images form an *atlas*, as this term is used in differential geometry.

5. **The interpretive form**: this unifies formerly incompatible images of reality by means of constructing a *model* of one system in the other. One essential tool is a *translation* between the categories in which the incompatible images

were originally formulated.[10]

6. ***The integrative form***: this brings resources for the unification of a larger number of mutually incompatible pictures of reality. It does this by separating the *structure* of a picture from its *ontological basis*. This form of language played a key role in the structuralist approach.

7. ***The constitutive form***: while the integrative form used a neutral ontological basis into which all structures could be embedded, the constitutive form produces the ontological constitution of objects, which was previously left to the ontological basis.

8. ***The conceptual form***: this makes explicit the possibilities of the ontological constitution.

A more detailed exposition of the theory of re-codings, with a number of examples from the history of synthetic geometry and of algebra, can be found in *Patterns of Change*. I chose to present at least an outline of that theory in order to show that structuralism perceives mathematics through the lens of one particular form of language – ***the integrative form***. This form is relatively late, having originated in the 19th century, and so enables a faithful reconstruction of a large part of contemporary mathematics. This makes structuralism plausible in the eyes of mathematicians and makes it one of the main positions in the philosophy of mathematics. It is not a position that would ascribe to mathematics features that are foreign to its practice. On the contrary, structuralism articulates aspects of mathematics that play an important role in its practice. The integrative form of language is sufficiently rich to make structuralism an interesting and important philosophical position.

3.1 Structuralism and the integrative form of language

Structuralism is a view of mathematics that is often connected with the creation of non-Euclidean geometry and abstract algebra. Both of these changes are relativizations connected to the integrative form of language. There exist, however, eight forms of the language of mathematics and the integrative form is just one of them. Mathematics had accumulated a long history before the structuralist approach to mathematics was established, and only on the basis of that approach does it make sense to interpret mathematical objects as places in structures. Thus the question arises of how to deal with the five forms of language preceding the integrative

---

[10]The interpretive form of language is the form in which non-Euclidean geometries were discovered.

form. We can maintain that before the emergence of structuralism, mathematicians studied structures, only without knowing.

What is more difficult, however, is the problem of how structuralism can deal with the later two forms. In principle we can say that structuralism identifies mathematics with its particular developmental phase. Before mathematics reached the structuralist level, it passed through a variety of simpler stages, and it is likely that in future its development will bring further forms, so Resnik's structuralism resembles Frege's logicism in that some particular historical form of mathematics (in Frege's case, a particular instrument of symbolic representation and in Resnik's case, a particular form of language) is identified with the whole of mathematics. In a previous paper I called Frege's logicism *linguistic pessimism*, that is, the view that mathematics has lost its creative power and will never create any new representational tools: it will thus remain stuck in the framework of Fregean logic through its entire future development. However, Frege's linguistic pessimism was unjustified. Mathematicians such as Gödel, Turing, Post, and Church created a new instrument of symbolic representation, by means of which mathematics surpassed the limits imposed by Frege's logicism.

Resnik's structuralism can be called *abstractive pessimism* i.e. the view that in the formation of structural mathematics, mathematics reached the maximal level of abstraction, and during its entire future development nothing essentially more abstract, nothing that would exceeded the boundaries of structural mathematics, would arise. Because Resnik refers to the mathematics of the period 1930-1970, when the structuralist approach culminated, we lack the advantage of sufficient hindsight that would allow us to find the next increases of abstraction that would break the limits into which structuralism locks mathematics. Yet in spite of this, I would like to formulate a *thesis of abstractive optimism*, according to which it is futile to try to enclose mathematics within a certain form of language, as Resnik tries to do.

3.2 The uniqueness of mathematical objects

Balaguer writes: "*One of the central ideas behind realism is supposed to be that our mathematical theories are about **unique** collections of abstract objects – e.g., arithmetic is supposed to be about **the** natural-number sequence – but when we look closely, it seems that realism is, in fact, incapable of delivering this result.*" Furthermore, he continues: "*Thus, even if mathematical objects are just positions in structures, and even if they are incomplete with respect to the properties they possess, it still seems to be very likely that there are multiple structures that exist outside space and time, that satisfy all of the desiderata for being the natural-number se-*

quence, and that differ from one another only in ways that no human being has ever thought about. Thus, I do not think it is at all obvious that structuralism can salvage unique structures for our mathematical theories to be about." ([2], pp. 116)

Geometry was confronted with something similar after the discovery of non-Euclidean geometry. In the interpretation of this discovery in ([17], pp. 124-133) it is shown that this discovery is related to the creation of a new form of language, which I called the *interpretative form*. In the framework of this form, mathematical reality is no longer only passively perceived, but an interpretative distance is introduced. It is this interpretive distance that allows us to think in different geometries. In the following *integrative form* an incorporation of the different geometries into one system occurred. Balaguer's objection can be interpreted as the identification of realism with *realism within the perspectivist forms of language*. Assumptions of this kind of realism are:

1. There is one *fully determined* world that is independent of us.

2. Our theories create *pictures* of that world.

3. The viewpoint from which a picture is produced *is neither a part of the world nor of its picture*.

The fact that during its history mathematics has created at least eight different forms of language clearly shows that it is not necessary to identify realism with realism within the perspectivist form of language. The *projective form of language* brought an incorporation of the viewpoint, from which the picture of reality is created, into the picture itself (thus changing point 3). This was crucial for the creation of the concept of geometric transformation. The *coordinative form of language* went further and replaced the picture of reality with its map (thus changing point 2). The *compositive form* replaced the map with an atlas (in the technical sense, as this term is used in differential geometry), and the *interpretive form* brought the idea of the pluralism of possible interpretations (thus changing point 1).

Balaguer's objection seems to be unjustified. It is his decision to identify realism with realism within the perspectivist form of language. That kind of realism cannot explain the practice of mathematics, which in the 17th century had already abandoned the perspectivist form of language: in projective geometry the *principle of duality* shows a remarkable symmetry of geometrical reality, which cannot be fully understood on the basis of the perspectivist form of language. The belief in the uniqueness of reality is a preconception which Resnik is not obliged to accept. The realistic interpretation of a certain area of human practice aims at interpreting as the effects of a certain ontological substrate the phenomena that we encounter within this practice. However, for an ontological interpretation of the constructive and deductive practice of mathematics we are not obliged to accept the thesis of the uniqueness of mathematical reality, because this practice abandoned that particular thesis long ago.

On the other hand, we must admit that from the point of view of the theory of relativizations, Resnik and Balaguer commit a similar mistake. Both raise features that are *adequate for one form language of mathematics* as binding for the ontological basis *of the whole of mathematics*. The only difference is that Balaguer took the first (perspectivist) form, which mathematics had long ago abandoned, whilst Resnik chose the sixth (integrative) form. Balaguer thus has problems with the interpretation of the practice of current mathematics, which long ago abandoned the assumption of the existence of a unique universe, whilst Resnik has problems with the interpretation of the practice of the early stages of mathematics, when the structural approach to mathematics did not yet exist.

4 Conclusion

A common objection to structuralism is that its three components – *the argument of indispensability*, *the postulationalist epistemology*, and *the structuralist ontology* are unrelated and actually they represent three independent positions. I believe that there is nothing wrong with this situation. I see Resnik's position as a complex philosophy of mathematics and Resnik simply chose the name of one of its components as the label of the whole position. In my view Resnik's position is remarkable in that it takes into account all three levels of the complexity of the language of mathematics (i.e. *idealization*, *re-coding*, and *relativization*). Resnik's structuralism reflects each of these three levels in a manner consistent with historical facts. In this respect it is quite different from many philosophies of mathematics, which take a principle valid for one of these three levels, and reduce the complex practice of mathematics to processes at this level. Against a background of such simplistic approaches, Resnik's position is very attractive. The juxtaposition of Resnik's theory with the history of mathematics has led at each of these three levels to similar results.

At the **level of idealizations** Resnik's theory can be made more plausible if we realize that instead of the *psychological theory of abstraction* we should base structuralism on the *linguistic theory of idealization*. Then, thanks to analogies between the Euclidean idealization, on which mathematics is based, and the Newtonian idealization, which is the basis of physics, Resnik's indispensability argument becomes more plausible and can be defended against many objections.

At the **level of re-codings**, which is the basis of the epistemological aspect of

Resnik's theory, we may realize that mathematical knowledge is gained by means of a particular instrumental practice. Thus the instruments, by means of which we acquire knowledge of the ideal mathematical objects, and these ideal objects themselves, are in a *conventional relation of rules of a language game* and not in a *factual relation* of some resemblance. This allows us to overcome objections concerning the causal inertness and non-localizability in space and time. The instruments themselves are physical objects and so they are causally active and localized in space and time. Thus Resnik's posits can be enriched by a theory of instruments of symbolic and iconic representation.

Only at the **level of relativizations** we encountered some limitations in Resnik's position. It is true that a large portion of contemporary mathematics can be plausibly interpreted as the study of structures. *Bourbaki's project* upon the foundations of mathematics succeeded to a large extent – the way how mathematicians today conduct, present, and teach mathematics is more or less compatible with the structuralist account. Nevertheless, from the point of view of a historian of mathematics, structuralism is only one of eight interpretations of the subject matter of mathematics. However, if we realize that Resnik's structuralism is based on *one of the forms of language*, we can generalize his position. Every account of the ontology of mathematics must be *based on a particular form of language*. Resnik has chosen one particular form of language for the whole of mathematics. The ontological account of mathematics would require that an interpretation of every segment of mathematical practice take into account the particular form of language upon which such a practice were based. Resnik developed an ontological account of a fragment of mathematics that is perhaps the most relevant one today, but his account must be *supplemented by seven further accounts*, one for each of the particular forms of language. Resnik simply wrote one of the eight possible books - or one of the eight chapters of a comprehensive philosophical account of mathematics - the one that is perhaps most interesting for the mathematicians of today. Resnik's position must *not* be seen as a negative thesis that claims no other mathematics except structuralist mathematics exists, but rather as a positive thesis, that his book is a model of how the accounts of mathematics based on the other forms of language should be developed. The other forms of language present not a refutation but rather complement Resnik's position. We have to try to articulate the philosophical assumptions of the remaining seven forms of language with the same precision and detail as Resnik has articulated the philosophical assumptions of the structuralist account.

Thus an ideal philosophy of mathematics would be one in which the realist aspiration of mathematics would be based on a **linguistic theory of idealization**; the epistemology of mathematics would be based on the interpretation of mathematics as a collection of language games constituted of various **instruments of symbolic**

and iconic representation and the ontological interpretation of the universe of objects made accessible by means of these representational instruments would be pluralistic, providing space for ***all eight forms of language***. Resnik has led us a long way down the road leading to this ideal and the fact that several elements of this ideal philosophy of mathematics can be found in his theory indicates that Resnik's efforts have carried us close to such a goal.

References

[1] Archimedes. The works of Archimedes. In *Great Books of the Western World*, volume 11, pages 403–592. Encyclopedia Britannica, Chicago 1952, 1952.

[2] M Balaguer. Michael Resnik. Mathematics as a science of patterns. Book review. *Philosophia Mathematica*, 7:108–126, 1999.

[3] P. Benacerraf. What numbers could not be. *Philosophical Review*, 74:47–73, 1965.

[4] P. Benacerraf. Mathematical truth. *Journal of Philosophy*, 70:661–680, 1973.

[5] E. Brieskorn and H. Knörrer. *Plane Algebraic Curves*. Birkhäuser Verlag, Basel, 1986.

[6] R. Descartes. *The Geometry of René Descartes*. Dover, New York, 1954.

[7] S. Drake, editor. *Discoveries and opinions of Galileo*. New York, Doubleday Company, 1957.

[8] Euclid. *The Thirteen Books of the Elements*. Dover, New York, 1956. Translated by Sir Thomas Heath.

[9] J. Folina. Michael Resnik. Mathematics as a science of patterns. Book review. *Notre Dame Journal of Formal Logic*, 40:455–472, 1999.

[10] G. Frege. Funktion und begriff. In G. Frege, editor, *Funktion, Begriff, Bedeutung*, pages 17–39. Göttingen, Vandenhoec & Ruprecht, 1989. English translation in: Translations from the Philosophical Writings of Gottlob Frege, edited by P. Geach and M. Black, Basil Blackwell, Oxford, 1952, pp. 21-42.

[11] G. Galilei. *The Assayer*. 1957. In [7].

[12] G. Hellman. *Structuralism*. 2005. In [34].

[13] E. Husserl. *Die Krisis der europäischen Wissenschaften und die transzendentale Phänomenologie*. Evanston, 1970. English translation The Crisis of European Sciences and Transcendental Phenomenology by David Carr.

[14] L. Kvasz. History of geometry and the development of the form of its language. *Synthese*, pages 141–186, 1998.

[15] L. Kvasz. Changes of language in the development of mathematics. *Philosophia mathematica*, 8:47–83, 2000.

[16] L. Kvasz. Galilean physics in light of husserlian phenomenology. *Philosophia Naturalis*, 39:209–233, 2002.

[17] L Kvasz. *Patterns of Change, Linguistic Innovations in the Development of Classical Mathematics*. Birkhäuser Verlag, Basel, 2008.

[18] L. Kvasz. Kant's philosophy of geometry – on the road to a final assessment. *Philosophia Mathematica*, 19:139–166, 2011.

[19] L. Kvasz. Galileo, Descartes, and Newton – founders of the language of physics. *Acta Physica Slovaca*, 62:519–614, 2012.

[20] L. Kvasz. Kuhn's structure of scientific revolutions between sociology and epistemology. *Studies in History and Philosophy of Science*, 46:78–84, 2014.

[21] N. Lobachevski. O nacalach geometrii (on the foundations of geomery, in russian). In *Polnoje sobranie socinenij (Collected works)*, volume 1. GITTL, Leningrad, 1946.

[22] B. Mandelbrot. *The Fractal Geometry of Nature*. W. H. Freeman and Co., New York, 1982.

[23] R. Netz. *The Shaping of Deduction in Greek Mathematics*. Cambridge, Cambridge University Press, 1999.

[24] S. Psillos. *Scientific Realism: How Science Tracks Truth*. Routledge, London, 1999.

[25] M. Resnik. Mathematics as a science of patterns: Ontology and reference. *Nous*, 15:529–550, 1981.

[26] M. Resnik. Mathematics as a science of patterns: Epistemology. *Nous*, 16:95–105, 1982.

[27] M. Resnik. Beliefs about mathematical objects. In A. D. Irvine, editor, *Physicalism in Mathematics*, pages 41–71. Kluwer Academic Publishers, Dordrecht, 1990.

[28] M. Resnik. Applying mathematics and the indispensability argument. In A. Ibara J. Echeverria and T. Morman, editors, *The Space of Mathematics*, pages 115–131. Walter de Gruyter, 1992.

[29] M. Resnik. What is structuralism? In D. Prawitz and D. Westerståhl, editors, *Logic and Philosophy of Science in Uppsala*, pages 355–364. Kluwer Academic Publishers, Dordrecht, 1994.

[30] M. Resnik. *Mathematics as a Science of Patterns*. Clarendon Press, Oxford, 1997.

[31] S. Shapiro. Mathematics and reality. *Philosophy of Science*, 50:523–548, 1983.

[32] S. Shapiro. Logic, ontology and mathematical practice. *Synthese*, 79:13–50, 1989.

[33] S. Shapiro. *Philosophy of Mathematics: Structure and Ontology*. Oxford University Press, Oxford, 1997.

[34] S. Shapiro, editor. *The Oxford Handbook of Philosophy of Mathematics and Logic*. Oxford University Press, New York, 2005.

[35] B. van der Waerden. *Moderne Algebra*, volume I. Springer Berlin, 1930.

[36] B. van der Waerden. *Moderne Algebra*, volume II. Springer Berlin, 1931.

[37] C. Weizsäcker. *Zum Weltbild der Physik*. S. Hirzel, Stuttgart, 13th edition, 1990.

A Complex Problem For Formalists

Arezoo Islami
Department of Philosophy, San Francisco State University
arezooi@sfsu.edu

Abstract

The *Formalist* conception of mathematics, which the physicist Eugene Wigner in "The Unreasonable Effectiveness of Mathematics in the Natural Sciences" defended, and eventually became the dominant philosophy of mathematics among the 20th century physicists has many shortcomings. It is this formalism that creates the *miracle* of the applicability of mathematics in modern physics and turns the relationship between mathematics and physics into a *happy accident* for which we don't have a "rational explanation". Through a historical reflection on complex numbers, their origin and development, I aim to show that this formalism does not hold up to scrutiny.

Keywords: Formalism, Unreasonable Effectiveness, Applicability of Mathematics, Complex Numbers

1 Introduction

In his seminal 1960 paper, the physicist Eugene Paul Wigner described the effectiveness of mathematics in physics as a *miracle*, which we neither *understand* nor *deserve*. In more than one place in "The Unreasonable Effectiveness of Mathematics in the Natural Sciences" he highlighted the "mysterious", "miraculous" and "puzzling" aspects of the applicability of mathematics in physics:

> [T]he enormous usefulness of mathematics in the natural sciences is something bordering on the mysterious and that there is no rational explanation for it. [27, p.223]

I thank Mark Steiner, Carl Posy, Ladislav Kvasz, Tom Donaldson, and Mic Detlefsen for their insightful suggestions. My special thanks are to Thomas Ryckman for his enormously useful comments on several revisions of the paper.

> It is difficult to avoid the impression that a miracle confronts us here, quite comparable in its striking nature to the miracle that the human mind can string a thousand arguments together without getting itself into contradictions, or to the two miracles of the existence of laws of nature and of the human mind's capacity to divine them. [27, p.229]

And finally he concluded by another striking statement:

> The miracle of the appropriateness of the language of mathematics for the formulation of the laws of physics is a wonderful gift which we neither understand nor deserve. We should be grateful for it and hope that it will remain valid in future research and that it will extend, for better or for worse, to our pleasure, even though perhaps also to our bafflement, to wide branches of learning. [27, p.237]

Pondering about the relationship between mathematics and physics is in no way unprecedented or unique to Wigner. Galileo famously wrote that the book of nature is written in the language of geometry:

> Philosophy [nature] is written in that great book which ever is open before our eyes – I mean the universe – but we cannot understand it if we do not first learn the language and grasp the symbols in which it is written. The book is written in mathematical language, and the symbols are triangles, circles and other geometrical figures, without whose help it is impossible to comprehend a single word of it; without which one wanders in vain through a dark labyrinth. [11]

And Kant asked how is the *synthetic a priori* possible. Yet it is Wigner's particular characterization of the problem that turns it into a miracle, a problem for which there is no rational explanation. Jesper Lützen in 2011 rightly argued that

> Before the twentieth century the effectiveness of mathematics does not seem to have been considered as a miracle. To be sure the mathematical nature of natural laws was often referred to in religious terms such as *God geometrizes*, but that was not considered miraculous. On the contrary, miracles were occurrences that seemed to defy the mathematical laws of nature.[19]

Wigner's formalist[1] conception of mathematics is one, and perhaps the most im-

---

[1]Formalism in philosophy of mathematics is the view that mathematics is no more than a formal game. Just as in a game of chess pieces are moved according to particular rules, symbols of mathematics are manipulated according to a set of rules. The symbols are arbitrary and uninterpreted otherwise. This view had gained popularity among the scientists of mid 20th century.

portant, aspect of his view that creates the *mystery* of applicability.[2] It is this formalism that problematizes the relationship between mathematics, the formal game, and physics, the study of nature, to an unprecedented degree. It creates miracles *out of the thin air* as Ferreirós dramatically put it [10], and turns the applicability of mathematics in the natural sciences into a *happy accident*. (Unger and Smolin [26], also Colyvan [4], Lützen [19], Grattan-Guinness [12], Longo [18]) More importantly, this conception of mathematics gives an inaccurate picture of how a concept is first developed and how it evolves in time.

Wigner defines mathematics as "the science of skillful operations with concepts and rules that are invented *only* for this purpose".[27] The emphasis is on invention (as opposed to discovery), and on invention *only* for the purpose of mathematics itself. On this formalist view, new mathematical concepts are invented based on intra-mathematical criteria such as formal beauty. For instance, Wigner wrote, the operations among the ordered pairs of numbers follow the same rules as operations among fractions. The operations that govern irrational numbers, or "sequences of numbers" are governed by the same rules as operations among rational numbers. And the list continues. Wigner's idea is that the more advanced concepts "were so devised that they are apt subjects on which the mathematician can demonstrate his ingenuity and sense of formal beauty". [27] That these concepts end up being applicable or even indispensable in the study of nature is what Wigner considers to be a *miracle*.

This formalist conception of mathematics, shared by many of Wigner's contemporaries, is a characteristic of the 20th century conception of abstraction, unique to what we can call *pure abstract mathematics* of the 20th century.

Paul Dirac in 1939 wrote:

> [T] he mathematician plays a game in which he himself invents the rules while the physicist plays a game in which the rules are provided by Nature, but as time goes on it becomes increasingly evident that the rules which the mathematician finds interesting are the same as those which Nature has chosen. [9, p.123]

David Hilbert in his 1919 lecture argued:

> We are confronted with the peculiar fact that matter seems to comply well and truly to the formalism of mathematics. There arises an unforeseen unison of being and thinking, which for the present we have to accept like a miracle.

---

[2]It has been argued that another source of this puzzle is Wigner's realism with respect to physics. Wigner's views of physics, on my view are best understood as anti-realist. For an elaborate discussion see Islami 2016 [15].

As an example of the mysterious and unreasonable effectiveness of mathematics in physics, Wigner highlights complex numbers which (on his view) were invented based on their formal beauty and not their applicability. Therefore, their ubiquitous use in physics, especially quantum mechanics seems to be inexplicable,

> The complex numbers provide a particularly striking example for the foregoing [the unreasonable effectiveness of mathematics]. Certainly, nothing in our experience suggests the introduction of these quantities. Indeed, if a mathematician is asked to justify his interest in complex numbers, he will point, with some indignation, to the many beautiful theorems in the theory of equations, of power series, and of analytic functions in general, which owe their origin to the introduction of complex numbers. The mathematician is not willing to give up his interest in these most beautiful accomplishments of his genius. [27, 225]

While Wigner is right about the original appearance of imaginaries (for the purpose of solving problems in mathematics itself), he is not right about their consequent use and development. Yet Wigner is not the only one misreading the history of imaginaries. The story of the development of complex numbers is often told with an air of mystery along these lines: the imaginaries were invented to solve quadratic equations in the 16th century. For centuries they were treated with suspicion and doubt as it seemed impossible for a negative number to have a square root. So they were called "nonsensical", "absurd", "impossible", "useless","sophistic", and later "imaginary"– existing only in mind. What is surprising is that these strange "numbers" ended up playing decisive roles in the 20th century physics such as quantum mechanics. They appear for instance in both Heisenberg's canonical commutation relation, and Schrödinger's wave equation. Hence, the argument goes, we have a miracle!

Christopher Pincock describes the situation as follows:

> They [complex numbers] were introduced into mathematics based on reflections on real numbers. Rounding out real numbers to include complex numbers led to a superior mathematical theory. The complex numbers lacked any intuitive physical significance, but in spite of this they provide a crucial part of mathematics deployed in science. The mystery, then, is that mathematics developed independently of applications should prove so effective when it is eventually applied. [23, p.17]

To call the applicability of mathematics a mystery or a miracle is quite unsatisfying and moreover, misleading. A contextual and historical study of the development of imaginaries presents us with a case against this kind of formalism and its subsequent adherence to miracles. The calculational power of imaginaries as solutions to the cubic equation in the 16th century later manifests itself in diverse fields of mathematics, connecting trigonometry to the study of exponential functions, vectors etc. While the introduction of imaginaries in the 16th century was faced with skepticism and resistance, and the mathematicians who decided to work with them adopted a more nominalist attitude, later a transition was made to a more realistic conception. What made complex numbers useful to the physicists of the 20th century wasn't their manipulability in Cardano's formula (in the 16th century) but their connection to trigonometric functions through Euler's identity and ultimately their use in the Fourier series.

The applicability of mathematics in physics is a rich and complex problem apt for a historically-sensitive philosophical investigation. This paper focuses on one such case: the applicability of complex numbers in quantum mechanics. Through a brief historical reflection, I hope to show the *reasonable* effectiveness of complex numbers in the original sources of quantum mechanics (i.e work of Heisenberg, Born, Jordan and Dirac) . As a result, I suggest to modify the historical narrative we often accept with regard to the origin and development of complex numbers from the 16th to the 20th century.

I begin with my solution to the problem of applicability of complex numbers in quantum mechanics based on the role of the Fourier series and Euler's formula in the following section. To understand what this solution amounts to we need to study the origin and the piecemeal development of imaginaries. It is through a brief historical investigation that we see how imaginaries came to have the calculational power that made them useful in the equations of quantum mechanics.

2 Complex Numbers in Quantum Mechanics

The starting point in the original sources of quantum mechanics, papers of Werner Heisenberg (1925), Max Born, Pascual Jordan (1925) and Paul Dirac (1926), is the use of the Fourier series to represent quantum theoretical entities.

Heisenberg started his paper [14] with a system described classically, where its time dependent position $x(t)$ was represented by a Fourier series as:

$$x(t) = \sum_{-\infty}^{\infty} a_\alpha e^{i\alpha\omega t}$$

And he argued that in quantum theory, Fourier coefficients must correspond to the transition from the state n to $n - \alpha$ and therefore the Fourier series, in the quantum case must be written as:

$$x(t) = \sum_{-\infty}^{\infty} a(n, n-\alpha) e^{i\alpha\omega(n,n-\alpha)t}$$

Note that already in the classical case, position is represented by a Fourier series, which as a result of Euler's formula ($e^{ix} = \cos x + i \sin x$), is written with exponential functions to the power i. In its quantum counterpart, it was precisely this exponential form that made the use of Ritz-Rydberg rule for frequency combination easy and straightforward. (For, based on Ritz-Rydberg rule $\omega(n, n-\beta) = \omega(n, n-\alpha) + \omega(n-\alpha, n-\beta)$, it can easily be seen that $e^{i\omega(n,n-\beta)t} = e^{i\omega(n,n-\alpha)t} e^{i\omega(n-\alpha,n-\beta)t}$)

In Born and Jordan paper[1], once again, the quantum theoretical entities of position p and momentum q are represented by their Fourier series as,

$$p = \sum_{\tau=-\infty}^{+\infty} p_\tau e^{2\pi i \nu \tau t}$$

$$q = \sum_{\tau=-\infty}^{+\infty} q_\tau e^{2\pi i \nu \tau t}$$

which in turn, in a few steps leads to the *canonical commutation relation*, which Born and Jordan call the "quantum condition":

$$[p, q] = \frac{h}{2\pi i}$$

In the Fourier series representations of the quantum theoretical entities of position p and momentum q, we already have the imaginary unit i. It is exactly this i that appears in the *canonical commutation relation*.

The derivation of the commutation relation is a straightforward mathematical result for any two non-commutating entities. Given Heisenberg's insight, based on experimental results [3], that quantum quantities of position and momentum don't commute, Born and Jordan carried out these calculations using matrices.

The general mathematical result is called the generalized uncertainty principle: that, for any pair of observables corresponding to operators that don't commute (i.e. "incompatible observables"), there is an uncertainty principle:

$$\sigma_A{}^2 \sigma_B{}^2 \leq (\frac{1}{2i} \langle [\hat{A}, \hat{B}] \rangle)^2$$

---

[3] See Alain Connes' interesting discussion of how Heisenberg came up with the idea of noncommutative entities. [5]

The proof of this principle is based on the Schwarz inequality[4] and properties of Hermitian operators (observables). (For more details, and a proof of this principle, see 3.5 in Griffiths's *Quantum Mechanics*[13, pp.110-112])

In his path to find the classical analogue to the commutator of non-commutative quantum theoretical entities, Dirac [8] also used their Fourier series representations. For any two dynamical variable x and y,

$$[x,y] = (xy - yx)(n,t) =$$
$$\sum_{\alpha,\beta} \{x(n, n-\alpha)y(n-\alpha, n-\alpha-\beta) - y(n, n-\beta)x(n-\beta, n-\alpha-\beta)t\}e^{i\omega(n,n-\alpha-\beta)t}$$

Using the quantization condition for the action variable, once again, he arrived at the conclusion that in the limit of large quantum numbers, $xy - yx$ corresponds to the Poisson bracket multiplied by a factor of $\frac{ih}{2\pi}$. The appearance of the imaginary unit i here is a result of substituting Fourier representations of quantum quantities in the Bohr-Sommerfeld condition.

What is fundamental to all these papers is the use of Fourier series for these quantities, in its exponential form. Given Euler's formula this form provides a powerful tool in carrying out calculations with much more ease than its trigonometric counterpart. Notice however, since all observables are at the end real quantities, in such calculations complex numbers only provide a tool for arriving at the results in the domain of real numbers. It is in this sense that as Hadamard said, "*The shortest path between two truths in the real domain passes through the complex domain.*"[17]

To understand exactly what my solution to the problem of applicability (of complex numbers in quantum mechanics) amounts to we first need to understand the definition and development of the Fourier series and Euler's formula. Moreover, it is not clear at the outset how these concepts have their origins in the introduction of imaginaries in the 16th century. A brief historical study of the evolution of complex numbers from their first appearance in the algebra of the 16th century to their use in the analysis of the 19th century is necessary for this purpose. The rest of this paper aims to present such historical reflection.

3 Roots of Negatives

Imaginary unit i, we are told, was first invented as a solution to the quadratic equation $x^2 + 1 = 0$. A common textbook reads along these lines:

---

[4]The Schwarz inequality is $|\langle f|g \rangle|^2 \geq \langle f|f \rangle \langle g|g \rangle$

> The real equation $x^2 + 1 = 0$ led to the invention of i (and also $-i$) in the first place. That was declared to be the solution and the case was closed. [24, p.330]

This is easy and straightforward, nonetheless inaccurate. The quadratic equations of this sort were considered impossible and not dealt with any further. In fact, unlike one tends to believe, mathematicians don't invent new numbers, or other mathematical entities on demand.[5] As we see throughout history, the introduction of new concepts is a complex process for a variety of mathematical and metaphysical reasons.[6]

In this case, what led to the introduction of these solutions in the 16th century were cubic equations which were known to have real positive solutions. They first appeared in Girolamo Cardano's (1501–1576) *Ars Magna* (*The Great Art* or *The Rules of Algebra*) in 1545 in his method for solving a cubic equation. [7] In the same book, chapter XXVII, Cardano raises this question, the solution of which also includes the roots of negative numbers: "If someone says to you, divide 10 by two parts, one of which multiplied into the other shall produce..40, it is evident that this case or question is impossible. Nevertheless we shall solve it in this way."[3][p.219] The answers he got, in modern notation, were $5 + \sqrt{-15}$ and $5 - \sqrt{-15}$, which involve imaginary numbers. Cardano, however, didn't pursue this further because he thought that such results are "useless" and "sophisticated":

> This is truly sophisticated, since with it one cannot carry out the operations one can in the case of a pure negative and other. ... For example, in this case, you could divide 10 into two parts whose product is 40; add 25, the square of one-half of 10, to 40, making 65; from the share root of this subtract 5 and also add 5 to it; you have parts with the likeness of $\sqrt{65} + 5$, $\sqrt{65} - 5$. While these numbers differ by 10 their sum is $\sqrt{260}$ not 10. So progresses arithmetic subtly, the end of which, as is said, is as refined as it is useless.[3][p.220]

---

[5] I have an uneasy feeling saying this. The standards for acceptance of new entities has changed drastically throughout the history. In particular the history of mathematics in the 20th century shows less resistance than preceding centuries. It is then that we deal with abstract pure mathematics which is independent from applicability and intuition. The questions the standards of acceptance in each era are interesting and require the space of their own. For now I focus solely on imaginaries.

[6] See Meir Buzaglo's *The Logic of Concept Expansion*[2] and Morris Kline's *Mathematical Thought from Ancient to Modern Times*[17] for an interesting discussion.

[7] Algebra came to Europe by way of the translation of Al-khwarizmi's *Algebra* (c. 825), in which he has used the method of *al-jabr* and *al-moqabala* in solution of problems including linear and quadratic equations. The italian Algebraists of the 12th century received the Latin translation of his book and expanded it to new problems. Cardano is famous for solving cubic and biquadratic equations. See van der Waerden [7], and Struik [25]

A few decades before Cardano, the general solution to the cubic equation was considered to be nearly impossible. Luca Pacioli (c. 1447–1517), at the end of his 1487 book (published in 1494) noted that the solution to the cubic equation is "as impossible at the present state of science as the quadrature of the circle." He stated that for equations in which:

numero, cosa e cuba (n, x, x^3)
or numero, cense e cubo (n, x^2, x^3)
or numero, cubo e censor de censor (n, x^3, x^4)

appear, "it has not been possible until now to form general rules". [7][p.47]

Historically, the quadrature of the circle was considered to be the measure of the difficulty of a mathematical problem. While Pacioli was right about the difficulty of quadrature of a circle –in fact it was proven in 1882 to be impossible– he was mistaken in his assessment of the cubic equation.

Within a decade, another Italian mathematician, Scipione del Ferro (1465–1526), discovered a solution to the *depressed cubic*, a case of the general cubic with the second degree term missing, that is, in modern notation, $x^3 + px = q$ (p, q are positive real numbers, or more accurately, magnitudes).[8] del Ferro's method was to express the solutions to the depressed cubic as sum of two terms: $x = u + v$. (The way that mathematicians at the time expressed the problems and the method of their solution was in words. There were symbols neither for the unknown, our x, nor for the coefficients (as we show them using a,b,c,..p,q). Moreover, the problems they solved were specific problems, with specific coefficients in the equations. And the method was not explained in general terms. (Thanks to Ladislav Kvasz and Carl Posy for drawing my attention to this important historical fact.)

In modern notation, the formula for solving the equation is,

$$x = \sqrt[3]{\frac{q}{2} + \sqrt{\frac{q^2}{4} + \frac{p^3}{27}}} - \sqrt[3]{-\frac{q}{2} + \sqrt{\frac{q^2}{4} + \frac{p^3}{27}}}$$

This looks too complicated to see how it relates to complex numbers. In an actual example, for the depressed cubic $x^3 + 6x = 20$ we have,

$$x = \sqrt[3]{10 + \sqrt{108}} - \sqrt[3]{-10 + \sqrt{108}}$$

which is nothing but $x_1 = 2$. Once we have this solution (which is actually a real number) to the depressed cubic $x^3 + 6x = 20$, we can easily find the other two

---

[8]The general cubic $x^3 + ax^2 + bx + c = 0$ can be reduced to this form by this substitution $y = x + 1/3a$.

through the quadratic formula:
$$x_2 = -1 + 3\sqrt{-1}$$
and
$$x_3 = -1 - 3\sqrt{-1}$$

Here you see the imaginary $\sqrt{-1}$, but this is not what del Ferro saw since finding *one* real *positive* answer to the cubic equation was all he was after (and therefore didn't substitute the real solution to find the rest).[9]

As the tradition in those days was, del Ferro kept his formula secret to use it in a public challenge, which actually never happened. However, he shared his discovery with a few of his students, one of whom, Antonio Maria Fior, claimed to have the solution and challenged the better known mathematician, Niccola Fontana, known as Tartaglia(c. 1499–1557) to a contest. This motivated Tartaglia to work fiercely on the problem and rediscover del Ferro's solution to the depressed cubic. On the day of the contest, Tartaglia succeeded to show his solution (to this problem as well as to 29 others) and won the contest.[10] Possessing the secret of the cubic formula, Tartaglia kept it to himself.

Cardano heard about this secret and was curious to know the solution to the cubic equation. So he and his assistant, Lodovico Ferrari (1522–1565), invited Tartaglia to Milan and insisted that Tartaglia shares the secret with them. While refused at first, Tartaglia finally shared his formulae with Cardano on the condition that it remains secret. Cardano was exposed to the answer once he took the oath of secrecy. Cardano writes:

> Scipio del Ferro of Bologna well-nigh thirty years ago discovered this rule and handed it on to Antonio Maria Fior of Venice, whose contest with Niccolo Tartaglia of Bresdica gave Niccolo occasion to discover it. He gave it to me in response to my entreaties, though withholding the demonstration. Armed with this assistance, I sought out its demonstration in forms. This was very difficult. [3][p. 96]

Nonetheless, it didn't quite remain a secret! Cardano saw the papers of del Ferro where this formula appeared and felt that he no longer has to keep his oath of

---

[9] del Ferro didn't publish any of his results. Cardano in his *Ars Magna* mentions him as having the solution. See chapter XI of *Ars Magna*.

[10] It is remarkable that once it becomes a known fact that the solution to a mathematical problem is possible, and in fact someone has found it, others rediscover it much easier than before. It is generally the case that when the time is ripe, there are multiple simultaneous discoveries in mathematics and sciences. As Kline puts it elegantly, "mathematical discoveries, like springtime violates in the woods, have their season which no human can hasten or retard."[17]

secrecy to Tartaglia. He seems to have rediscovered Tartaglia's solution for himself and started to work on the general case of the cubic equation.

The way that Cardano expressed the problems and solutions is quite different from ours. For instance he expressed the equation in this way: "A cube and unknowns are equal to a number" (*Cubus et res equals numero*. Our x, is the unknown, which he used the Italian word *cosa* or the Latin word, *res* to refer to. The word number, *numero* is used to refer to numerical coefficients. See, for instance, Cardano chapter XI in *Ars Magna*. It is in Descartes' work that the notation we use today is introduced.

Although, Cardano called these solutions *sophisticated*, he still worked with them. *Using* them wasn't the problem for him. The puzzle was that they appeared in the formula for the cubic that had only positive *real* solutions.

Take this example:
$$x^3 = 15x + 4$$
Here, $p = 15$, $q = 4$ and $\frac{q^2}{4} = 4 < \frac{p^3}{27} = 125$ which means we have negative roots in Cardano's formula. However, when we calculate all the roots, $x = 4$, $x = -2 + \sqrt{3}$ and $x = -2 - \sqrt{3}$, we realize that all three are *real* solutions. What seemed paradoxical to Cardano was that the real root $x = 4$ in his formula is expressed in terms of imaginaries. Morris Kline writes:

> One would think that the fact that real numbers can be expressed as combinations of complex numbers would have caused Cardano to take complex numbers seriously, but it didn't.[17, p.266]

Cardano didn't take them seriously but the fact that he wrote them down gave them, "a symbolic existence".[6] [11]

What seemed to have confused Cardano was his lack of understanding of complex conjugates, as we can easily see today. It is this insight into complex conjugates that Rafael Bombelli (1526–1572), thirty years after Cardano's *Ars Magna* calls his "wild thought": "It was a wild thought, in the judgment of man, and I too was for a long time of the same opinion. The whole matter seemed to rest on sophistry rather than on truth. Yet I sought so long, until I actually proved this to be the case."[17, p.712]

Bombelli's *wild thought* was that since, in modern notation, $2 + \sqrt{-121}$ and $2 - \sqrt{-121}$ differ only in a sign, this must might be true of their cubic roots. [12]

---

[11] It is important to ask why he didn't take them seriously. Mathematics for Cardano and his contemporaries was properly understood as the study of magnitudes, arithmetic or geometrical.

[12] So he assumed that there are reals a, b such that: $\sqrt[3]{2 + \sqrt{-121}} = a + \sqrt{-b}$ and $\sqrt[3]{2 - \sqrt{-121}} = a - \sqrt{-b}$ With a few step of calculation, he deduced that $a = 2$ and $b = 1$. Therefore, $\sqrt[3]{2 + \sqrt{-121}} + \sqrt[3]{2 - \sqrt{-121}} = (2 + \sqrt{-1}) + (2 - \sqrt{-1}) = 4$

Bombelli went further, introduced a notation for what we call i (namely *piu di meno*), and $-i$ *meno di meno* and developed some rules for imaginaries, in modern notation, as: $(-1)i = -i$, $(+i)(+i) = -1$, $(-i)(+i) = +1$, $(-i)(-i) = +1,\ldots$. His work was the beginning of an arithmetic for the imaginaries. However, an acceptable meaning or interpretation for these solutions was still missing.

The situation wasn't much better for the negatives. In the 16th century, negatives were still considered to be absurd. Cardano and many others thought of them as "fictitious", mere symbols and formal tools to obtain real solutions. Descartes for instance, called negative solutions to equations "false" solutions since "they were representing something less than nothing". [17, pp.251-259] Real solutions only included *magnitudes*, consisting of what we call positive rational and positive irrational numbers, which we obtain geometrically. To the Italian mathematicians, the term *number/ numero* was to be used only for coefficients or constants in equations, which were positive integers. Magnitudes were the only acceptable solutions to the equations they solved. [13]

4 Geometrical Interpretation

Following the Greek tradition, what mathematicians required for an entity to be considered legitimate, that is as a magnitude, was a geometrical representation. In fact, following Euclid, all "algebraic" demonstrations from Al-Khwarizmi to Cardano were followed by geometric demonstrations.[14] Mathematicians of the 17th century, René Descartes (1596–1650) and John Wallis (1616–1703) in particular, tried to come up with geometrical interpretation for the *imaginaries* but neither succeeded in going very far: it was not until a century later that the way to understand imaginaries geometrically became clear.

In another front, as early as 1629, the mathematician Albert Girard (1595–1632) claimed that all equations of algebra have as many solutions as the degree of the equation (a result that later became known as the *Fundamental Theorem of Algebra*). This claim was rather *vague* and *unclear*. It didn't specify the domain to which these solutions belong.

In his *L'invention nouvelle en l'algébre* Girard gave examples of the equations with complex solutions. For instance in modern notation for the equation $x^4 = 4x - 3$

---

[13]Throughout the paper I have used the term "number" quite liberally, mostly in our modern sense of the word. For a more historically accurate account, we need to be careful about the transition from magnitudes to numbers which happened in the 19th century. Given that my goal is to describe the development of the "use" of imaginaries, I hope the interchangeable use of imaginaries and complex numbers doesn't create much confusion.

[14]See [7] for an interesting discussion of the development of Algebra.

we have four solutions:
$$1, -1, -1+\sqrt{-2}, -1-\sqrt{-2}$$
For him, once again, these solutions were impossible:

> We must therefore always remember to keep this in mind: if someone were to ask what is the purpose of the solutions that are impossible, then I answer in three ways: for the certitude of the general rule, and the fact that there are no other solutions, and for its use. The use is easy to see, since it serves for the invention of solutions of similar equations as we can see in Stevin's arithmetic, in the 5th difference of the 71st problemï£¡ [25][p.86]

Yet Girard recognized their usefulness in solving equations (for instance cubic equations as the reference to Simon Stevin's Arithmetic indicates). René Descartes continued the work on the fundamental theorem of algebra. He formulated it in 1637, in *Géométrie* as,

> Know then that in every equation there are as many distinct roots, that is the values of the unknown quantity, as is the number of dimensions of the unknown quantity.[7]

To Descartes we owe mainly our current algebraic notation, x,y,.. for the unknown, a,b,c,.. for coefficients,. Also he introduced terms "real", "imaginary" (for he thought we can "imagine" that there be n such roots).[15] Descartes attempted to find a geometrical representation for these roots but didn't go too far.

Casper Wessel (1745–1818) in his 1797 paper, "On the Analytic Representation of Direction: An Attempt" found a way to the geometrical representation of complex numbers. His paper is focused on representing vectors (directed line segments) and the operations over them geometrically. The horizontal axis is the usual x-axis with the unit 1 and the vertical axis is considered as the axis of imaginaries with the unit $i = \sqrt{-1}$. In his paper a vector $\overrightarrow{OP}$ is represented by $a + b\sqrt{-1}$ where a is on the x-axis and b on the y-axis.

For every two vectors $\overrightarrow{OP}$ and $\overrightarrow{OQ}$, he defined their sum by positioning the second vector's initial point at the terminal point of the first vector and the resulting sum is a vector which connects the starting point of the first vector to the terminal point of the second (constructing a triangle).[16] Wallis arrived at this result by

---

[15] Descartes was after a *Universal Science* and his analytic geometry which applied the "Algebra of the Moderns" to the "Geometry of the Ancients" was an example of such universal science. The introduction of notation replacing letters for words, which started by Francois Viete and mainly completed by Descartes, made the formulation of general rules in this universal science possible.

[16] This method is equivalent to calculating the sum as the diagonal of the parallelogram with the two vectors as its adjacent sides.

an analogy from parallel line segments (vectors in the same direction), which is calculated by placing one at the terminal point of the other and adding their length together.

Wessel's discoveries were result of his generalization of these operations with real numbers. For instance, he observed that if the product of two real number 3, 5 is 15 then $\frac{15}{3} = \frac{5}{1}$. So assuming that we have a unit vector, analogously we can say that the product of two vectors obeys the same rule: that is its length to one vector is equal with the length of the other vector to one. (Descartes used a similar rule with the product of line segments in his *Geometry*.[17]) The product of vectors is a new vector, in modern notation, $\overrightarrow{OS}$ whose length is such that $|\overrightarrow{OS}|$ to $|\overrightarrow{OP}|$ is equal to $|\overrightarrow{OQ}|$ to 1 (and it's angle to x-axis is the sum of the angles of $\overrightarrow{OP}$ and $\overrightarrow{OQ}$ to the x-axis). About direction of this product, by the same analogy, he thought it should differ from each of the vectors by the same factor that each of these vectors differ from the unit vector. That is if $\overrightarrow{OP}$ makes the angle α with x-axis, and $\overrightarrow{OQ}$ the angle β then the product $\overrightarrow{OS}$ must make the angle $\alpha + \beta$ with the unit vector on the x-axis.

Wessel's paper provided the geometric meaning of $\sqrt{-1}$ in this way. Suppose that $\sqrt{-1}$ can be represented by a vector whose length is l and whose angle with the real axis is α (which is shown as $\sqrt{-1} = l\angle\alpha$). Using Wessel's vector multiplication rule, we can calculate the product of that vector with itself. The result will have the length of l^2 and the angle of 2α. That is $(\sqrt{-1})^2 = -1 = l^2\angle 2\alpha$. We know that $-1 = 1\angle 180°$, so $l^2\angle 2\alpha = 1\angle 180°$, and therefore, $l^2 = 1$ and $2\alpha = 180°$ which means $\alpha = 90°$. What this important discovery says is that

$$\sqrt{-1} = 1\angle 90°$$

What this means is that by rotating the unit vector on the x-axis by 90° counterclockwise, we get to the imaginary unit, which is $\sqrt{-1}$. This is the property of the imaginary unit that makes it also a "rotation operator".

Wessel's paper, which contained these remarkable breakthroughs on the geometrical meaning of the imaginaries went unfortunately unnoticed until 1897 when it was translated from Danish to French and republished.

Meanwhile, Jean-Robert Argand (1768–1822), a self-taught Swiss bookkeeper came up with a different geometrical interpretation in 1806. He used his understanding of negatives, as positives with a different direction, to give an interpretation of the square root of negative one. He asked himself about how to turn $+1$ into -1. He then noticed that if he rotates the unit line segment 90° counterclockwise and then

---

[17] The French text *La Géométrie* has been published, together with the English translation by David Eugene Smith and Marcia Latham in 1954.

repeat this rotation he gets from $+1$ to -1. But this is exactly what happens if we multiply $+1$ by $\sqrt{-1}$ twice. From this he concluded that we can think of $\sqrt{-1}$ and the counterclockwise 90° rotation as one and the very same thing. In the same way, $-\sqrt{-1}$ is the clockwise 90° rotation.[18]

Argand's work fell on a similar fate as Wessel's; it attracted very little attention. However, two decades after Argand published his work, in 1828, John Warren(1796–1852) published a book in Cambridge which included a rather complete presentation of complex numbers to that day titled, *A Treatise on the Geometrical Representation of the Square Roots of Negative Quantities*.[19] William Rowan Hamilton (1805–1865) read this book a year later and started his work on complex numbers. For him it was not acceptable that $\sqrt{-1}$ hasn't been given a purely algebraic meaning but was only expressed geometrically. He wrote (much later in 1853),

> I .. felt dissatisfied with any view that should give to [imaginaries] from the outset a clear interpretation and *meaning*; and wished that this should be done, for square roots of negatives, without introducing considerations so *expressly geometrical* as those which involved the conception of an angle.[22]

Influenced by Kant, Hamilton believed since geometry is the science of space, algebra should be the science of time. So he thought of associating $\sqrt{-1}$ with time. In 1843, he presented a paper in which he defined complex numbers in terms of ordered pairs, with operations of addition and multiplication defined in this way: for every two ordered pairs (a, b) and (c, d) we have,

$$(a, b) + (c, d) = (a + c, b + d)$$
$$(a, b) \times (c, d) = (ac - bd, bc + ad)$$

These algebraic definitions of course weren't arbitrary, as Hamilton admitted, since he used his understanding of complex numbers (which was given geometrically prior to him) to come up with these definitions. According to Hamilton's notation, every complex number $a+bi$ could be written as the ordered pair (a, b). The latter notation was advantageous to him, because it avoided the "absurd" $\sqrt{-1}$. Then every real number r, can be written as $(r, 0)$. It is obvious that with a few steps of calculation

---

[18] As Nahin argues, Argand's 1806 discovery was simultaneous with another discovery by Bué. [21, pp.75-77]

[19] Note that he still refers to negatives as *negative quantities*. The definition of number had to be liberated before it could include negatives as well as positives, and imageries as well as reals.

(using both notations), we have[20]

$$(0,1) = \sqrt{-1} = i$$

In another front, Carl Friedrich Gauss (1777–1855) gave the first proof of the fundamental theorem of algebra in 1799 in his doctoral dissertation. In another proof in 1849 he used imaginaries, claiming that it is no longer necessary to avoid them. He wrote: the "intuitive meaning of complex numbers in [their] geometrical representation completely established and more is not needed to admit these quantities into the domain of arithmetic." [17, p.631][21]

Unlike Wessel and Argand, Gauss didn't represent complex numbers geometrically as vectors but as points on the complex plane, written as $a + ib$. The addition and multiplication of these points were defined geometrically, just as Wessel and Argand had done.[22] Gauss, moreover, is the first mathematician to use the letter i for the imaginary unit.

What led to the acceptance of these entities arguably was a shift in the proper subject of mathematics, which gradually moved from magnitudes to the more abstract idea of numbers and then structures.[23] Yet what is important to their development is that they were used. As van der Waerden argues after Descartes the leading mathematicians made free use of these numbers [7, p.175] and that is what is crucial to their usefulness in other areas of mathematics and physics.

---

[20]Using modern notion of a field we have: the set of complex numbers $\mathbb{C}$ with the operations of addition and multiplication, as Hamilton defines them, is an algebraic field. That is, $(\mathbb{C}, +)$ is an abelian group with the identity element $(0,0)$, $(\mathbb{C} - (0,0), \times)$ is an abelian group with the identity element $(1,0)$, and the operation of $\times$ is distributive over $+$. The filed of $\mathbb{R}$ is a subfield of $\mathbb{C}$.

[21]Gauss later in 1831 argued the terminology was partially responsible for the long and slow progress that was made with these numbers. He writes, "If this subject has hitherto been considered from the wrong viewpoint and this enveloped in mystery and surrounded by darkness, it is largely an unsuitable terminology which should be blamed. Had $+1$, -1 and $\sqrt{-1}$, instead of being called positive, negative and imaginary (or worse still impossible) unity, been given the names, say, of direct, inverse and lateral unity, there could hardly have been any scope for such obscurity." This is not quite accurate however. Mathematics had to liberate itself from magnitudes as the its only proper subject of study to allow for imaginaries to be accepted in the domain of "arithmetic". To argue for this in details requires a space of its own.

[22]The complex plane is referred to sometimes as the Gaussian plane or the Argand plane.

[23]See Ernest Nagel's interesting paper titled "Impossible Numbers" for an argument to this conclusion. There he argues that the acceptance of these "impossible numbers" as Nagel calls them, marks an important shift in our conception of mathematics: it frees mathematics from its ties to arithmetic.[20] Thanks to Mark Steiner for the reference to Nagel.

5 Complex numbers in the 18th century

As a consequence of the body of work on problems from physics in the 17th century, the concept of a function and instances of some simpler functions such as the $log(x)$ (which Euler wrote as lx) and e^x were known. In his important book on analysis, *Introductio in analysin infinitorum*, Leonhard Euler (1707–1783) defined a function as:

> A function of a variable quantity is an analytical expression of some kind composed from that variable quantity and from constant numbers or magnitudes.

This definition of function is due to his teacher Johann Bernouli (1667–1748) who first gave a formal definition of a function (in 1718). By an analytical expression, Euler meant expressions that can be formed using only algebraic operations such as addition, subtraction, multiplication, division, power, root and their composition. He further divided functions into algebraic and transcendental functions and devoted parts of the book to the better study of the latter (a goal that he declared in the beginning of ch. 4). Examples of transcendental functions are exponentials and logarithms which he studied through their expansions.

Using the method of power series expansion, in chapter 7 he defined exponential and logarithmic functions as

$$e^x = 1 + \frac{x}{1} + \frac{x^2}{1.2} + \frac{x^3}{1.2.3} + \frac{x^4}{1.2.3.4} + etc.$$

$$\log(1+x) = x - \frac{x^2}{2} + \frac{x^3}{3} - \frac{x^4}{4} + etc.$$

Moreover, by then the trigonometric quantities were also systematically studied. The sin and cos of sum and differences of two angles were known, for instance, due to the works of mathematicians including Johann Bernouli. Euler in 1739 published a paper about harmonic oscillators in which he used the *sine* as a solution the differential equation.

Following his work on trigonometric functions and power series expansion, in chapter 8 of *Introductio*, he gave the power series expansion of *sine* and *cosine*:

$$sinx = x - \frac{x^3}{1.2.3} + \frac{x^5}{1.2.3.4.5} - \frac{x^7}{1.2.3.4.5.6.7} + etc.$$

$$cosx = 1 - \frac{x^2}{1.2} + \frac{x^4}{1.2.3.4} - \frac{x^6}{1.2.3.4.5.6} + etc.$$

A heated controversy of that time was over the log of negatives and imaginaries. Bernouli argued that $\log(-1) = 0$ while Leibniz strongly disagreed. Bernouli's argument was along these lines. Since

$$\frac{d(-x)}{-x} = \frac{dx}{x}$$

so using integration $\log(-x) = \log x$. So $\log(-1) = 0$ since $\log 1 = 0$.

Leibniz rejected this argument because

$$d(\log x) = \frac{dx}{x}$$

only for $x > 0$. If Bernouli was right it would mean that $\log \sqrt{-1} = 0$ since

$$\log \sqrt{-1} = \frac{1}{2} \log(-1)$$

The key clarification in the calculation of the logarithm of imaginaries came as a result of the work of Roger Cotes(1682 –1716) in 1714. His result, in modern notation, was:

$$\sqrt{-1}\phi = \log_e(\cos \phi + \sqrt{-1} \sin \phi)$$

Euler rediscovered this result in 1748 using a different route. In a letter to Bernouli in 1740 Euler noted that the differential equation

$$\frac{d^2y}{dx^2} + y = 0$$

where $y(0) = 2$ and $\frac{dy}{dx}(0) = 0$, has two solutions:

$$y = 2\cos x$$

$$y = e^{\sqrt{-1}x} + e^{-\sqrt{-1}x}$$

He thought they must be equal and therefore, he declared in 1743 that:

$$\cos x = \frac{e^{\sqrt{-1}x} + e^{-\sqrt{-1}x}}{2}$$

$$\sin x = \frac{e^{\sqrt{-1}x} - e^{-\sqrt{-1}x}}{2}$$

From which he concluded Cotes result which is now universally known as Euler's formula:

$$e^{ix} = \cos x + i \sin x$$

This outstanding formula connects two apparently unconnected areas of mathematics together– trigonometric functions and imaginary quantities.[24]

The expressions in parenthesis are nothing but the power series expansion of *sin* and *cos*. Therefore using the power series expansions, once again, we arrive at Euler's formula.

6 Euler's formula

Euler's formula is one of the most remarkable formula in mathematics in part because it connects complex numbers with trigonometry and in part because it has many interesting consequences.

Let's look at some of the more simple results. For the case of $x = \pi$ we have the famous "Euler's identity" $e^{i\pi} = -1$ or $e^{i\pi} + 1 = 0$. This identity is especially interesting as it involves the addition, multiplication, exponentiation as well as five important mathematical numbers 1, 0, e, π and i. Among many important properties that these numbers have we can enumerate the following. 0 and 1 are the identity elements for addition and multiplication respectively. e, Euler's number, is a positive irrational number which is the base of the natural logarithm. π is another positive irrational number which is the ratio of the circumference of a circle to its diameter (in radian it is the angle measure of 180°.) About i is the imaginary unit, about which we have said a lot. This is the identity that Richard Feynman famously called "our jewel" and "the most remarkable formula in mathematics".

As a consequence of Euler's identity, we can calculate the logarithm of negative numbers.
$$e^{i\pi} = -1$$
so,
$$ln(-1) = i\pi$$

For $x = \frac{\pi}{2}$ we have $e^{i\pi/2} = i$. As a result we can calculate the logarithm of imaginary numbers. We have:
$$ln(i) = \frac{i\pi}{2}$$
or
$$\pi = \frac{2}{i} ln(i)$$

---

[24]Euler's confidence in this formula was due to his knowledge of the power series expansion of e^x, *sin* and *cos*. In modern notation we have: $e^x = 1 + x + \frac{1}{2!}x^2 + \frac{1}{3!}x^3 + \cdots$ which for ix is:
$e^{ix} = 1 + (ix) + \frac{1}{2!}(ix)^2 + \frac{1}{3!}(ix)^3 + \cdots$ Rearranging the real and imaginary parts we have,
$e^{ix} = (1 - \frac{1}{2!}x^2 + \frac{1}{4!}x^4 + \cdots) + i(x - 13!x^3 + 15!x^5 + \cdots)$

Now if we raise both sides to the power i we have,

$$i^i = (e^{i\pi/2})^i = e^{i^2\pi/2} = e^{-\pi/2} = 0.2078¡£ \cdot$$

It is remarkable that an imaginary number to the power of an imaginary number is a real number! It is in fact, a positive irrational number. What is more remarkable is that i^i has infinitely many real values, $e^{-\pi/2}$ is just one.

Another breakthrough that came with Euler's identity was in the context of trigonometric identities. From Euler's formula we can write trigonometric functions sin and cos in terms of the exponential function. [25]

As another example to see the power of using complex numbers, we can look at how Euler calculated the following integrals:[26]

$$I_1 = \int_0^\infty \sin(s^2)ds$$

and

$$I_2 = \int_0^\infty \cos(s^2)ds$$

These integrals appeared in the analysis of the physics of a coiled spring, also in certain hydrodynamical questions and later in the study of the diffraction of light (and therefore they are famous as Fresnel integrals). These integrals occupied Euler's mind for long without much success. He wrote in 1743, " We must admit that analysis will make no small gain should anyone find a method whereby approximately ay least the value of [these intgrals] would be determinedï£¡. This problem does not seem to be unworthy of the best strength of geometers."

While in 1743 he found power series expansions that would help with finding the numerical value of these integrals, Euler didn't come up with that "method" until 1781, when he used complex quantities in his calculation. (See Appendix)

---

[25] If $e^{i\theta} = \cos(\theta) + i\sin(\theta)$, we have: $\cos(\theta) = \frac{e^{i\theta}+e^{-i\theta}}{2}$ and $\sin(\theta) = \frac{e^{i\theta}-e^{-i\theta}}{2i}$ Having these expressions is very helpful in calculating of the product of trigonometric functions, since working with exponential function is so much easier than calculating with trigonometric functions.
$\sin(\alpha)\cos(\beta) = (\frac{e^{i\alpha}-e^{-i\alpha}}{2i})(\frac{e^{i\beta}+e^{-i\beta}}{2}) = \frac{e^{(\alpha+\beta)}-e^{-i(\alpha+\beta)}+e^{i(\alpha-\beta)}-e^{-i(\alpha-\beta)}}{4i} = \frac{2i\sin(\alpha+\beta)+2i\sin(\alpha-\beta)}{4i} = 1/2\sin(\alpha+\beta) + 1/2\sin(\alpha-\beta)$
which is just the familiar rule for product of trigonometric functions. The same approach can be used to calculate other products such as $\cos(\alpha)\cos(\beta)$ and $\sin(\alpha)\sin(\beta)$.

[26] See Algebraic Analysis in the 18th century in *A History of Analysis* by Jahnke [16] for an elaborate discussion.

7 Fourier Series

Closely connected to Euler's formula is one of the most remarkable mathematical results: *Fourier series*. In 1807, Joseph Fourier (1768-1830) presented his groundbreaking idea that every function can be represented by a trigonometric series satisfying certain conditions to the Academy of Science in Paris.[27] The idea of a trigonometric series was introduced before Fourier's birth as a response to the controversy in how to define a *function* (in works of Euler, Johann Bernoulli, Daniel Bernoulli, D' Alembert and their contemporaries). While for some mathematicians such as D' Alembert a function could only involve ordinary algebraic operations, Euler suggested a broader definition. For Euler a function f can be defined if we can draw its curve– which means the derivative of the function has to exist nearly at every point (if a function has a finite number of points without derivative we can still draw its curve. Think of $f(t) = |t|$.) Fourier's response, to jump ahead a bit, was that a function is defined when we can write its Fourier series.

The problem that led to Fourier's work was a problem in physics, motivated by musical instruments with strings such as violin.[28] The question is how does a perfectly elastic string, of uniform mass density along its length, with both ends fixed, move when is set in motion? In other words what is the wave equation for such vibrating string? The equation that D' Alembert came up with in 1747 for a one-dimensional string was:[29]

$$\frac{\partial^2 y}{\partial x^2} = \frac{1}{c^2}\frac{\partial^2 y}{\partial t^2}$$

The question is how to solve this equation given boundary conditions that the string is fixed on both sides (at the origin 0 and its end l):

$$(1) y(0,t) = 0$$

$$(2) y(l,t) = 0$$

and the initial conditions that the string is at rest before the deflection:

$$(3) \frac{\partial y}{\partial t}|_{t=0} = 0$$

---

[27] My discussion here is based on chapter 4 of Jahnke *A History of Analysis* [16], chapter 16 of Kline's [17], Langer's *Fourier's Series: The Genesis and Evolution of a Theory* and Nahin's book [21]

[28] Mathematics of the 18th century was unprecedented in its technicality. However, as Kline writes, "it was guided not by sharp mathematical thinking but by intuitive and physical insights."[17, p.400]

[29] The derivation is rather straightforward but since it is tangential to my discussion of Fourier series, I won't go into further details.

$$(4) y(x,0) = f(x)$$

The solution that Johann Bernoulli offered to this equation subject to the boundary and initial conditions was a trigonometric series:

$$y(x,t) = \sum_{n=1}^{\infty} c_n \sin(\frac{n\pi}{l}x) \cos(\frac{n\pi c}{l}t)$$

With the initial condition 4 we have,

$$f(x) = \sum_{n=1}^{\infty} c_n \sin(\frac{n\pi}{l}x)$$

So the problem now is finding these coefficients c_n. This is a particularly interesting result as it says the initial arbitrary deflection function $f(x)$ of a string is the sum of infinitely many *odd, periodic* functions. (By a periodic function, we mean a function that repeats itself. Mathematically speaking, a periodic function $f(t)$ is a function such that for a $T > 0$, $f(t+T) = f(t)$ for every $t \in D_f$. T, the period of the function, is the smallest positive number satisfying this condition. *sin* and *cos* are the most common examples of periodic functions.)

Fourier's work was also based on a problem from physics, the problem of finding the so-called *heat equation*. The question is about the diffusion or propagation of heat (with time) through a thin rod, with thermal conductivity. Just like the wave equation, in the one dimensional form we have:

$$\frac{\partial^2 u}{\partial x^2} = \frac{1}{k}\frac{\partial^2 u}{\partial t^2}$$

where $u(x,t)$ is the temperature along the rod and k is a constant. Fourier's insight was that by starting from a function $f(t)$ over the interval (a,b) we can define the function (1) over any interval $(0, L)$ (starting from 0) with a simple rule $\frac{L}{b-a}(t-a) \to t$ (2) over the interval $(-L, L)$ with an odd or even extension (the even extension of the function defines the function as symmetric with respect to the y-axis, whereas the odd extension is symmetric with respect to the origin (0,0))– and this is the original insight of Fourier. Finally the function defined over the interval $(-L, L)$ can be extended to $(-\infty, +\infty)$ as a periodic function with the period $2L$. This periodic function, Fourier claimed can be written with a trigonometric series.

$$f(t) = a_0 + \sum_{k=1}^{\infty}(a_k \cos(k\omega_0 t) + b_k \sin(k\omega_0 t))$$

where a_k and b_k are constant and $\omega_0 = 2\pi/T$. Recall T was the period of a function. ω_0 is what is called fundamental frequency of a function.

Using Euler's identity, as we saw in the previous section, we have:

$$\cos(k\omega_0 t) = \frac{e^{ik\omega_0 t} + e^{-ik\omega_0 t}}{2}$$

and,

$$\sin(k\omega_0 t) = \frac{e^{ik\omega_0 t} - e^{-ik\omega_0 t}}{2}$$

Using these in Fourier's series we have,

$$f(t) = a_0 + \sum_{k=1}^{\infty} \{a_k \frac{e^{ik\omega_0 t} + e^{-ik\omega_0 t}}{2} + b_k \frac{e^{ik\omega_0 t} - e^{-ik\omega_0 t}}{2}\}$$

$$= a_0 + \sum_{k=1}^{\infty} (a_k/2 + b_k/2) e^{ik\omega_0 t} + \sum_{k=1}^{\infty} (a_k/2 - b_k/2) e^{-ik\omega_0 t}$$

Taking the sum from $-\infty$ to ∞ we have:

$$f(t) = \sum_{k=-\infty}^{\infty} c_k e^{ik\omega_0 t}$$

where c_k's are constants and $\omega_0 = 2\pi/T$.[30]

Using Euler's formula enabled us to arrive at a representation of the Fourier series that is so much easier to deal with, in calculations (think of multiplication, addition, power, etc.) than the original trigonometric series. It is therefore a great tool in carrying out rather complex manipulations, and in this resides its calculational power.

8 Concluding Remarks

Based on the brief historical remarks, we can conclude that the piecemeal development of complex numbers was not led solely by formal beauty, *a la* Wigner or by reflections on real numbers *a la* Pincock.[31] There were a multitude of factors involved, among which we can mention the desire of mathematicians to solve problems in mathematics as well as problems originated in the sciences. Mathematical Analysis of the 18th and the 19th century from which we studied brief moments was

---

[30] The Fourier coefficients are $c_n = 1/T \int_T f(t) e^{-in\omega_0 t} dt$. I will not get into the derivation of these coefficients.

[31] Perhaps this notion of formal beauty is also anachronistic. Mid 20th century welcomes such emphasis on formal aspects of mathematics, which itself is a result of a change in our conception of what mathematics is.

an attempt to solve problems that were originated in modern physics, following the great scientific revolution of the 17th century.

Moreover, attempts to find geometrical representations for these quantities, and to give them as a result a more legitimate status as magnitudes, led mathematicians to a deeper understanding of the relationship between real and complex numbers (and later to understanding their algebraic structure, in Hamilton's work). Representing complex numbers on the Argand plane, as points or vectors, provided a basis for drawing on geometrical intuitions, which in turn led to connecting complex numbers to our experience of nature (with two dimensional plane and rotation operators). And, it is precisely in this sense that Wigner's characterization of these numbers doesn't hold up to scrutiny. (Recall Wigner's quote once again: "Certainly, nothing in our experience suggests the introduction of these quantities. Indeed, if a mathematician is asked to justify his interest in complex numbers, he will point, with some indignation, to the many beautiful theorems in the theory of equations, of power series, and of analytic functions in general, which owe their origin to the introduction of complex numbers. The mathematician is not willing to give up his interest in these most beautiful accomplishments of his genius."[27])

As a result of an extensive body of work on these numbers, mathematicians understood many of their important properties including what is expressed by Euler's formula connecting trigonometric functions and complex numbers. Euler's formula provided the basis for carrying out calculations with much more ease, since working with the exponential function proved to be much easier than dealing with trigonometric functions. Euler's formula, in turn, used in the Fourier series provided a powerful tool in calculations with functions. In particular, as I argued in the beginning, it allowed the quantum physicists to represent quantum theoretical entities and derive the canonical commutation relations.

The case of complex numbers is brought to provide *one* example of the *reasonable* applicability of specific mathematical concepts in physics. To solve the applicability problem for specific concepts we need to work on a case by case basis. In this process, on my view, attention to the details of the historical development of these concepts is of great importance, it is perhaps indispensable (as I hope to have shown in the case of complex numbers.)

A Appendix

Euler's method was based on what we today call the *gamma function* defined as:

$$\Gamma(n) = \int_0^\infty e^{-x} x^{n-1} dx, n > 0$$

For instance, $\Gamma(1) = 1$ and we can easily show that

$$\Gamma(n+1) = n\Gamma(n) \qquad (1)$$

Moreover for every positive integer n we have,[32]

$$\Gamma(2) = 1.\Gamma(1) = 1!$$
$$\Gamma(3) = 1.\Gamma(2) = 2!$$
$$\Gamma(4) = 1.\Gamma(3) = 3!$$
$$\ldots$$
$$\Gamma(n) = 1.\Gamma(n-1) = (n-1)!$$

Now for $n = \frac{1}{2}$ we have

$$\Gamma(\frac{1}{2}) = \int_0^\infty \frac{e^{-x}}{\sqrt{x}} dx$$

The integral looks rather difficult but it can be evaluated using the following trick. Let's change the variable $x = t^2$, then $dx = 2tdt$ and we have,

$$\Gamma(\frac{1}{2}) = \int_0^\infty \frac{e^{-x}}{\sqrt{x}} dx = \int_0^\infty \frac{e^{-t^2}}{t} dt = 2\int_0^\infty e^{-t^2} dt$$

So all we need to evaluate now is ,

$$I = \int_0^\infty e^{-t^2} dt \qquad (2)$$

Substituting t with arbitrary variables u and v, we have,

$$I^2 = (\int_0^\infty e^{-u^2} du)(\int_0^\infty e^{-v^2} dv) = \int_0^\infty \int_0^\infty e^{-(u^2+v^2)} dudv$$

The double integration is over $0 \leq u, v < \infty$ which is in the first quadrant of $u - v$ Cartesian coordinate, whereas in the polar coordinate where $u = r\cos\theta i$ and $v = r\sin\theta$ the integration is over $0 \leq r < \infty$ and $0 \leq \theta < \pi/2$. So for $r = u^2 + v^2$ we have,

$$I^2 = \int_0^{\pi/2} \int_0^\infty e^{-r^2} rd\theta dr d\theta$$

---

[32] Note that we can use the recursion relation in 1 to calculate $\Gamma(n)$ for negative ns. We have $\Gamma(n) = \frac{1}{n}\Gamma(n+1)$. Using a few steps we have $\Gamma(-\frac{1}{2}) = \frac{1}{-1/2}\Gamma(\frac{1}{2}) = -2\sqrt{\pi}$.

With a simple calculation we have

$$I^2 = \int_0^{\pi/2} \frac{1}{2} = \pi/4$$

So,

$$I = \frac{\sqrt{\pi}}{2}$$

Since in 2, $\Gamma(1/2) = 2I$, we have $\Gamma(1/2) = \sqrt{\pi}$

Euler's next step in evaluating I_1 and I_2 was to extend the gamma function to the complex plane. For the complex variable $u = \frac{x}{p+iq}$ where p and q are positive real variables we have,

$$\Gamma(n) = (p+iq)^n \int_0^\infty x^{n-1} e^{-px} e^{-iqx}$$

Using Euler's identity and using the polar form $p + iq = r(cos\alpha + i\sin\alpha)$, we have:

$$\int_0^\infty x^{n-1} e^{-px} \cos(qx) dx = \frac{\Gamma n}{r^n} \cos(n\alpha)$$

$$\int_0^\infty x^{n-1} e^{-px} \sin(qx) dx = \frac{\Gamma n}{r^n} \sin(n\alpha)$$

Now for $n = 1/2$, $p = 0$ and $q = 1$, we have,

$$\int_0^\infty \frac{\cos x}{\sqrt{x}} = \Gamma(1/2) \cos \pi/4 = \sqrt{\pi/2}$$

$$\int_0^\infty \frac{\sin x}{\sqrt{x}} = \Gamma(1/2) \sin \pi/4 = \sqrt{\pi/2}$$

Changing the variable $x = s^2$, we get to the original integrals:

$$I_1 = I_2 = \frac{1}{2}\sqrt{\pi/2}$$

Euler's remarkable trick using the gamma function is a method we use frequently in solving similar integrals. It is just that method that is as he put it, "not unworthy of the best strength of the geometers."

References

[1] Max Born and Pascual Jordan. On quantum mechanics. In Van der Waerden, editor, *Sources of Quantum Mechanics*, pages 277–306. Dover, 1925.

[2] Meir Buzaglo. *The Logic of Concept Expansion*. Cambridge University Press, Cambridge, 2002.

[3] Girolamo Cardano. *Ars Magna or The Rules of Algebra*. Dover Publications, New York, 1993.

[4] Mark Colyvan. Mathematics and the world. *Philosophy of Mathematics*, 2009.

[5] Alain Connes. *Noncommutative Geometry*. Academic Press, 1994.

[6] Tobias Dantzig and Joseph Mazur. *Number: The Language of Science*. Plume, 2007.

[7] B. L. Van der Waerden, editor. *A History of Algebra*. Springer Verlog, Berlin Heidelberg, 1985.

[8] Paul A. M. Dirac. The fundamental equations of quantum mechanics. In van der Waerden, editor, *Sources of Quantum Mechanics*, pages 307–320. Dover, 1925.

[9] Paul A. M. Dirac. The relation between mathematics and physics. *Proceedings of the Royal Society*, 59:122–129, 1939.

[10] José Ferreirós. Wigner's 'unreasonable effectiveness' in context. *Mathematical Intelligencer*, 2016.

[11] Galileo Galilei. *The Assayer*. Doubleday and Co., New York, 1623.

[12] Ivor Grattan-Guiness. Solving wigner's mystery: The reasonable (though perhaps limited) effectiveness of mathematics in the natural sciences. *Mathematical Intelligencer*, 30:7–17, 2008.

[13] David J. Griffiths. *Introduction to Quantum Mechanics (2nd edition)*. Benjamin Cummings, US, 2004.

[14] Werner Heisenberg. Quantum-theoretical reinterpretation of kinematic and mechanical relations. In van der Waerden, editor, *Sources of Quantum Mechanics*, pages 261–276. Dover, 1925.

[15] Arezoo Islami. A match not made in heaven: On the applicability of mathematics in physics. *Synthese*, 87:1–23, 2016.

[16] Hans N. Jahnke, editor. *A History of Analysis*. American Mathematical Society, United States, 2003.

[17] Morris Kline. *Mathematical Thought from Ancient to Modern Times*. Oxford University Press, 1972.

[18] Giuseppe Longo. The reasonable effectiveness of mathematics and its cognitive roots. *Geometries of Nature, Living Systems and Human Cognition*, pages 351–382, 2005.

[19] Jesper Lützen. The physical origin of physically useful mathematics. *Interdisciplinary Science Reviews*, 36:229–43, 2011.

[20] Ernest Nagel. Impossible numbers: A chapter in the history of modern logic. In *Teleology Revisited and Other Essays in the Philosophy and History of Science*, pages 166–194. Columbia University Press, New York, 1979.

[21] Paul J. Nahin. *An Imaginary Tale*. Princeton University Press, Princeton, New Jersey, 1998.

[22] John O'Neill. Formalism, hamilton and complex numbers. *Studies in History and Philosophy of Science*, 17:351–372, 1986.

[23] Christopher Pincock. *Mathematics and Scientific Representation*. Oxford University Press, 2014.

[24] Gilbert Strang. *Introduction to Applied Mathematics*. Wesley-Cambridge Press, 1986.

[25] D. J. Struik. *A Source Book in Mathematics*. Princeton, 1986.

[26] Roberto Mangabeira Unger and Lee Smolin. *The Singular Universe and the Reality of Time*. Cambridge University Press, 2014.

[27] Eugene P. Wigner. The unreasonable effectiveness of mathematics in the natural sciences. In *Symmetries and Reflections*, pages 222–237. Ox Bow Press, 1960.

Formalism and Set Theoretic Truth

Michael Gabbay
Department of Philosophy, Cambridge

Abstract

In this paper I extend an earlier formalist truth theory for Arithmetic to cover set theory. In effect, I propose that the truths of the language of set theory are the theorems of the axioms of ZFC extended with an ω-rule. However, I show that this truth definition can be presented without reference to infinitary proof rules and remains within the spirit of Hilbert's original finitism.

Keywords: Formalism, Set Theory, Arithmetic, Truth, Recursive Ordinal, Tree, Omega-Rule

1

In [2] I proposed an account of arithmetic truth which I argued is a formalist one. The proposal is not formalist in the traditional sense of regarding Arithmetic as the closure of some deductive system. In fact, the proposal shares much with the simple realist treatment of the basic arithmetic language: it presents a truth theory for it.

To exemplify the proposal consider the following language $\mathcal{L}_A$. The *objectual* terms are given by

$$t ::= x_1, x_2 \cdots \mid f(t_1 \ldots t_n)$$

The *arithmetic terms* are given by

$$\mathbf{t} ::= \mathbf{x_1}, \mathbf{x_2} \cdots \mid \mathbf{t_1} + \mathbf{t_2} \mid \mathbf{t_1} \times \mathbf{t_2} \mid 0 \mid 1$$

Among the arithmetic terms are *numerals* which are given by:

$$\mathbf{n} ::= 0 \mid \mathbf{n} + 1$$

The formulae are given by

$$A ::= a_1{=}a_2 \mid F(a_1 \ldots a_n) \mid \neg A \mid A_1 \wedge A_2 \mid \forall x A \mid \mathbf{a_1}{=}\mathbf{a_2} \mid \forall \mathbf{x} A \mid \exists^{\mathbf{t}} x A \mid \exists^{\mathbf{t}} \mathbf{x} A$$

On the level of atomic formulae $\mathcal{L}_A$ is strictly divided into an objectual fragment and an arithmetic fragment (identity either takes two objectual terms, or two arithmetic terms). The language also has the familiar truth functional connectives and quantifiers that bind objectual variables, and quantifiers that bind arithmetic variables; thus far the language is the same as that of a simple two sorted logic. The quantifiers $\exists^{\mathbf{t}}\mathbf{x}$ and $\exists^{\mathbf{s}}x$ bind the variables $\mathbf{x}$ and x respectively, but do not bind in $\mathbf{t}$. So for example in the formula

$$\exists^{\mathbf{t}}\mathbf{x}\exists^{\mathbf{s}}xA$$

the variables of $\mathbf{t}$ are *not* bound, but the instances of $\mathbf{x}$ in $\mathbf{s}$ *are* bound (by the outer quantifier $\exists^{\mathbf{t}}\mathbf{x}$), moreover both x and $\mathbf{x}$ are bound in A.

Let an *interpretation* I and a valuation v on that interpretation interpret the objectual terms and predicates in the familiar way.[1] Let O be a recursive collection of sentences including the atomic fragment of Peano Arithmetic formulated using the arithmetical fragment of the language above. So $O \vdash A$ or $O \vdash \neg A$ for every closed atomic arithmetic A. O may or may not derive some open atomic formulae.

We now give a truth theory for this language. Given an I and an O, we define what it is for A to be true under a valuation v written $v \vDash A$. For readability I present the clauses in three groups. First, first-order connectives.

$$\begin{array}{lll}
v \vDash Ft_1\ldots t_n & \text{iff} & v(t_1)\ldots v(t_n) \text{ satisfy } F^I \\
v \vDash \mathbf{n}=\mathbf{m} & \text{iff} & O \vdash \mathbf{n}=\mathbf{m} \\
v \vDash \neg A & \text{iff} & v \nvDash A \\
v \vDash \forall x A & \text{iff} & v(x \mapsto d) \vDash A \text{ for every } d \text{ (of the domain of } I) \\
v \vDash \forall \mathbf{x} A & \text{iff} & v \vDash A[\mathbf{x}/\mathbf{n}] \text{ for every } \mathbf{n}
\end{array}$$

Second the clauses for quantifying into objectual contexts.

$$\begin{array}{lll}
v \vDash \exists^{\mathbf{0}}xA \text{ is true} & \text{iff} & v(x \mapsto d) \nvDash A \text{ for any } d \\
v \vDash \exists^{\mathbf{n+1}}xA \text{ is true} & \text{iff} & v(x \mapsto d) \vDash A \text{ and } v(x \mapsto d) \vDash \exists^{\mathbf{n}}y(A \wedge x \neq y) \text{ for some } d
\end{array}$$

$$v \vDash \exists^{\mathbf{t}}xA \text{ is true} \quad \text{iff} \quad v \vDash \exists^{\mathbf{n}}xA \text{ where } O \vdash \mathbf{n}=\mathbf{t}$$

Finally the clauses for quantifying into arithmetic contexts.

$$\begin{array}{lll}
v \vDash \exists^{\mathbf{0}}\mathbf{x}A \text{ is true} & \text{iff} & v \nvDash A[\mathbf{x}/\mathbf{n}] \text{ for any } \mathbf{n} \\
v \vDash \exists^{\mathbf{n+1}}\mathbf{x}A \text{ is true} & \text{iff} & v \vDash A[\mathbf{x}/\mathbf{m}] \text{ and } v \vDash \exists^{\mathbf{n}}\mathbf{x}(A \wedge \mathbf{x} \neq \mathbf{m}) \text{ for some } \mathbf{m} \\
v \vDash \exists^{\mathbf{t}}\mathbf{x}A \text{ is true} & \text{iff} & v \vDash \exists^{\mathbf{n}}\mathbf{x}A \text{ where } O \vdash \mathbf{n}=\mathbf{t}
\end{array}$$

---

[1] I supplies a domain and a relation on that domain to each F, v assigns an element of the domain to each objectual variable x.

We can say that A is true simplicier, $\vDash A$ if A is true for any I, v and O.

We sacrifice elegance for simplicity having two types of quantifier, objectual and arithmetic, rather than one that can do either.

Intuitively, and abusing notation a little: $\forall x A$ means that every individual is A; $\forall \mathbf{x} A$ means that every number is A; $\exists^{\mathbf{t}} x A$ means that there are exactly $\mathbf{t}$ individuals that are A; and $\exists^{\mathbf{t}} \mathbf{x} A$ means there are exactly $\mathbf{t}$ numbers that are A.

The definition of truth for $\mathcal{L}_A$ is perfectly recursive. The interpretation and valuation determines the meaning or truth value of, say, $F t_1 \ldots t_n$. In the case of $\mathbf{t}_1 = \mathbf{t}_2$ the truth value depends on derivability O. The truth of an objectual quantifier depends on a familiar objectual interpretation. The truth an arithmetic quantifier depends on a substitutional interpretation.

The idea is that we obtain a truth theory for this language making use only of properties of the syntax. Nowhere are numerical terms required to refer to anything, or even do anything other than feature in derivations of atomic formulae or as substitution instances of quantifiers. We can then view this arithmetic language as a kind of logic. Numbers are a kind of logical constant (numerals) in that they have no denotation but nonetheless contribute to the truth conditions of sentences that contain them.[2] Arithmetic language looks and behaves like a referential language, but actually it isn't. Notice that the proposal is not dependent on any particular arithmetic language (i.e. any particular numeral system). The numerals are used rather than referenced to characterise the quantifier $\forall \mathbf{x}$. Had we used different notation, we would have described the same truth theory, just in different terms.

The benefits of such a view is that many debates depending on the apparent referential nature of an arithmetic language are bypassed. This is because we can replace talk of what number words refer to with talk of what contribution they make to the truth conditions of sentences containing them. And we can characterise that without regarding them as denoting expressions. For example, just as a debate about the indispensability (or reality) of conjunction is uninteresting, so is a debate about the indispensability or reality of Zero. Arithmetic becomes an exercise in expressive power: we use the arithmetic language it to express more propositions in more ways (so we can come to know them more easily).[3]

The proposal is thus formalist in the sense that it makes no attempt to analyse the concept of number as some kind of truth-maker for arithmetic truths. Instead it focuses on providing a account, ultimately dependent on a formal theory, of how arithmetic terms contribute to the truth conditions of sentences containing them,

---

[2] For example we need not see conjunction $\wedge$ as denoting anything, but merely as making a certain (truth functional contribution to any sentence containing it.

[3] The language above has the expressive power of a first order language with an additional 'there are finitely many...' quantifier.

including natural language sentences such as 'there are n Fs'.

2

The question I wish to begin to address here is how to extend this approach to higher mathematics, especially as it may appear in natural language. Not only is an account of Analysis desirable, but also an account of sets and other properties captured by set theory. For example

> The numbers are linearly ordered
> Some critics admire only each other
> The length of this rod in meters is transcendental

The truth conditions of all of these can be expressed in terms of sets, so a formalist account of set theory along the same lines as above would be ideal. We cannot play the same trick by choosing some suitable set theory, such as ZFC, and using a substitutional quantifier over closed terms. We would require a recursive and complete theory S of atomic formulae which we can use to build a truth theory for a set theory (or the real numbers) by induction on the complexity of its language, $\mathcal{L}_{\text{ZFC}}$ (using substitutional quantification to handle the quantifiers). But if we had such an S, then, since the language of set theory can express the recursive functions we could describe S in L, via Gödel coding. Then, since substitution is a recursive syntactic operation, we can describe the set of all true sentences of the truth definition as the closure of S under the clauses of the truth definition. But then the language would have its own truth predicate, which is impossible by Tarski's theorem. This argument applies not merely to the language of set theory but also a suitably rich language of real numbers, as each real number can stand in for a set of natural numbers. So a complete truth theory for set theory or even an elementary language of real numbers is impossible.

So it natural to treat set theoretic truth, not in terms of a complete truth theory, but as a kind of axiomatic extension of a complete truth theory such as the one above for Arithmetic. To explain exactly what this could mean is the purpose of this paper.

3

I shall turn to 'extending', in a sense, the truth theory of $\mathcal{L}_A$ with an axiomatisation of ZFC and propose this as for $\mathcal{L}_{\text{ZFC}}$. Actually, since we are not wedded to any particular notation for numerals, or to any particular accompanying O, we may simplify matters by using (any one of) set theory's own implementation of numbers as the basis for describing the truth conditions of arithmetic assertions.

So rather than extending the arithmetic fragment of $\mathcal{L}_A$, let us use its obvious implementation in $\mathcal{L}_{ZFC}$. So for example, we take the numerals to be terms for the finite ordinals and use $\forall x(x \in \omega \to \ldots)$ (for short we can write this as $\forall x_{\in \omega} \ldots$) in place of $\forall \mathbf{x}$. In place of the numerical quantification $\exists^{\mathbf{t}} \mathbf{x} A$ we use the formalisation in $\mathcal{L}_{ZFC}$ of

> there is a bijection from set $\mathbf{t}'$ to $\{x \mid A'\}$

where $\mathbf{t}'$ and A' are the implementations of $\mathbf{t}$ and A in $\mathcal{L}_{ZFC}$. We can now work entirely in $\mathcal{L}_{ZFC}$ (regarding $\mathcal{L}_A$ as a fragment of it).

We require a formalist truth definition for $\mathcal{L}_{ZFC}$ which depends on derivations using the axioms of ZFC, and derivations from ZFC to feed back into this truth definition. In short, what we want is something like this:

> A is a truth of ZFC when A is derivable from some true premises using the axioms of ZFC or A is $\forall x_{\in \omega} B$ and $B[x/\mathbf{n}]$ is true for each $\mathbf{n}$.[4]

This is circular and has the inconsistent theory as a solution.

Better definitions are that A is a true of ZFC when:

1. A is a member of the least set s that is closed under the the axioms of ZFC, Modus Ponens, and the condition that $\forall x_{\in \omega} B \in s$ if $B[x/\mathbf{n}] \in s$ for all $\mathbf{n}$ (i.e. the ω-rule).

2. A is a formula derivable using the ω-rule and the axioms of ZFC.

Both definitions validate this truth condition:

$$\forall x_{\in \omega} A \text{ is true iff } A[x/\mathbf{n}] \text{ is true for all } \mathbf{n}$$

Which is what we are looking for. But as they stand neither will do as a formalist account of truth in ZFC.

1. The first explicitly assumes language of sets, together with a truth definition, explicitly.

2. The second assumes a language of sets implicitly in two respects:

    (a) stating the ω-rule as an inference rule requires reference to the infinite set of premises.

---

[4]Here $\mathbf{n}$ is a closed term for an ordinal in ω.

(b) the "-able" of derivable suggests an inductive definition of derivations which, given the infinitary nature of the ω-rule requires induction on transfinite ordinals.

This is because derivations are defined inductively on their depth, so the present notion of derivation is defined something like this:

- An axiom A of ZFC is a derivation is a derivation of A.
- If D_1 and D_2 are derivations of A and $A \to B$ then
$$\frac{D_1 \quad D_2}{B}$$
is a derivation of B
- If $D_0, D_1 \ldots$ are derivations of $A[x/0], A[x/1] \ldots$ then
$$\frac{D_0, D_1 \ldots}{\forall x_{\in \omega} A}$$
is a derivation of $\forall x_{\in \omega} A$.

This is a legitimate inductive definition only if we have a measure of the complexity (here depth) of a derivation such that the complexity strictly increases when one of the two inference rules is applied. But suppose we apply the ω-rule where the depth of each premise D_n is n. Then the complexity of the derivation as a whole must be greater than n for each n: we need to assign it a transfinite ordinal ω. Generalise on this and we find we must define derivations by induction on the first ordinal that cannot be expressed as a countable limit of smaller ordinals (i.e. has co-finality greater than ω). This is the first uncountable ordinal ω_1. Thus a substantial amount of set theory is assumed in order to legitimise the proposed definition of set theoretic truth.

However we can find something very similar to 2 that is far more acceptable to a formalist.

4

The computable ω-rule is exactly the same as the ω-rule
$$\frac{A[x/\mathbf{0}] \quad A[x/\mathbf{1}] \quad \ldots}{\forall x_{\in \omega} A}$$
except it requires there to be a computable function that generates a derivation of $A[x/\mathbf{n}]$ for each $\mathbf{n}$. So now we may say that A is true of ZFC when

3. A is a formula derivable using the *recursive* ω-rule and the axioms of ZFC.

This takes care of (2a) above, the premise need not be regarded as an infinite set but a description of a computable function. But we have yet to deal with (2b): transfinite ordinals are required for the inductive definition of the infinitary proof system underlying the extension of the term 'derivable'.

What is interesting about the use of the ω-rule is that we feel we understand what it means without noticing its complexity. In general, we can grasp an inductive definition without thinking of it as an induction on some measure. A detailed mathematical characterisation of the definition, as an induction on some ordered structure, can plausibly seen as an explanation rather than a prerequisite for us to understand the definition. Perhaps this is what is meant by a comment of Hilbert and Bernays when discussing inductive definitions in their initial discussion of recursive definitions:

> Again, definition by recursion is not an independent principle of definition: Rather, within the framework of elementary number theory, recursion merely constitutes a convention for abbreviating descriptions of certain formation processes by which a numeral is obtained from one or more given numerals. [3, p27]

Similarly, within the general framework of derivation trees we can view the inductive definition as a means of computing the derivation trees, as long as all the inference rules are computable. Derivations using the computable ω-rule can be computable in the sense that a transcendental number can be computable: although we can never produce the whole entity, we have a computable function that will produce any finite part. In the case of a real number the finite part is an initial segment of its expansion, in the present case, as we shall see, the finite part is an ending segment of the derivation tree.

5

If $>$ is a binary relation then

- $s_1 \geq s_2$ means that $s_1 > s_2 \vee s_1 = s_2$,
- $s \in >$ means that $\exists s'(s' > s \vee s > s')$
- s is an *endpoint* of $>$ means that $s \in >$ and $\neg \exists s'(s > s')$
- $>$ and $>'$ are *disjoint* when $\neg \exists s(s \in > \wedge s \in >')$

Some binary relations are *trees*. A binary relation is a tree according to the following definition:

- For two distinct individuals, s_1, s_2, the relation $>$ such that $s_1 > s_2$ is a tree.

- Let $>$ be a tree and let s be an endpoint of $>$. Let f be a computable function f such that $f(n)$ is either 0 or is a tree, $>_n$, such that the $>_n$ are disjoint from each other and from $>$. Then relation $>'$ such that $s_1 >' s_2$ iff either

  - $s_1 > s_2$
  - $s_1 \geq s$ and $s_2 \in >_n$ for some n
  - $s_1 >_n s_2$ for some n

  is a tree.

Notice that on this definition all trees are transitive binary relations.

The intuitive description of this definition is that the simplest tree is a relation between two distinct elements and trees are expanded (downwards) by adding finitely or denumerably many branches to the bottom of endpoints (preserving transitivity) provided that it is computable what those branches are.

I claim that this definition of trees is perfectly intelligible, and can be used to characterise which binary relations are trees. We are describing a property of binary relations and saying that any relation that has that property is a tree.[5]

---

[5] We can present this definition in a more concrete way using the full spread of sequence symbols. Sequence symbols are defined by the syntax

$$\sigma ::= 0 \mid \sigma.n$$

where n is any numeral. A tree is then a property of sequences (identifying the sequences which describe paths in the tree).

Some preliminary definitions:

- $\sigma \succ \sigma'$ when σ is a substring of σ'.
- For any P, if $P\sigma$ and $\neg P\sigma.n$ for any n, then say σ is an *endpoint* in P.
- For any P, if either
  1. $\sigma = 0$ and $P\sigma$, or
  2. $\sigma = \sigma'.n$ and $P\sigma$ but $\neg P\sigma'$

  then say σ is a *startpoint* in P.

Now we define trees T as properties of sequences. To use familiar notation, we write $\sigma \in T$ for $T\sigma$.

- The property '... is σ' is a tree, for any σ.
- If T is a tree with endpoint σ, and f is a computable function such that, for each n, $f(n)$ is a tree T_n with startpoint $\sigma.n$ or is 0. Then the property T' such that $T'\sigma$ iff

6

An $\forall$-*derivation* is a standard first-order derivation witnessing either that

$$\forall x_{\in\omega} B \vdash_{\text{ZFC}} A \quad \text{or} \quad \vdash_{\text{ZFC}} A.$$

A *sequent* Γ is a finite list of formulae indexed by a numeral:

$$A_1^{n_1}, \ldots, A_k^{n_k}$$

Here A_i is a formula and n_i is a numeral indexing it. So, for example, if A, B and C are formulae then

$$B^{21}, A^0, C^{95}$$

is a sequent. Here, B^{21} does not represent the 21st of a list of Bs, it represents the formula B with a superscript 21.

Assume, via a suitable Gödel numbering that we have a (computable) ordering of the $\forall$-derivations. Now define a relation $>$ on sequents as follows.

(ω) $A^n, \Gamma > \Gamma, B[x/\mathbf{m}]^0, A^{n+1}$ for all $\mathbf{m}$ if the nth $\forall$-derivation is a derivation that $\forall x_{\in\omega} B \vdash_{\text{ZFC}} A$.

(S) $A^n, \Gamma > \Gamma, A^{n+1}$ if the nth $\forall$-derivation is *neither* a derivation that $\vdash_{\text{ZFC}} A$ nor that $\forall x_{\in\omega} B \vdash_{\text{ZFC}} A$ for any B.[6]

An alternative way of representing $>$ is using the horizontal line typically used to describe derivations. Here we place Γ below Γ' to represent that $\Gamma > \Gamma'$.

$$\frac{\Gamma, B[x/\mathbf{0}]^0, A^{n+1} \quad \Gamma, B[x/\mathbf{1}]^0, A^{n+1} \quad \Gamma, B[x/\mathbf{2}]^0, A^{n+1} \quad \cdots}{A^n, \Gamma} \; (\omega)$$

- $T\sigma$, or
- $f(n) \neq 0$ and $T_n \sigma$.

is a tree.

A tree the collection of its nodes σ and is transitively ordered by $\succ$. If σ is a node in a tree then if $\sigma.n$ is also a node then it represents (the start of) a branch from σ. Each σ may have finitely or infinitely many branches. In the infinite case, the branching is determined by a computable function, i.e. there is a computable function which determines whether or not there is a node $\sigma.n$ in the tree.

This notion of a tree can then be extended to handle relations between individuals in general. Say a T is *dressed* with S if there is a computable function that assigns a symbol $s \in S$ to each node σ of the T. So a tree of individuals S becomes a tree dressed with elements of S.

---

[6]In other words, if the nth $\forall$-derivation is either not a derivation of A at all, or it is a derivation of A from premises other than a single premise $\forall x_{\in\omega} B$ for some B.

$$\frac{\Gamma, A^{n+1}}{A^n, \Gamma} \ (S)$$

With the following side conditions:

(ω) the nth $\forall$-derivation witnesses $\forall x_{\in \omega} B \vdash_{\text{ZFC}} A$.

(S) (ω) is not applicable and the nth $\forall$-derivation is *not* a derivation that $\vdash_{\text{ZFC}} A$.

Described this way, $>$ or its transitive closure $>^*$ appears to be the reverse derivation relation. We can even make it look more like the definition of a derivation relation by adding a (redundant) clause

$$\frac{}{A^n, \Gamma} \ (Z)$$

with the side condition that the nth $\forall$-derivation witnesses $\vdash_{\text{ZFC}} A$ (i.e. by saying that no sequent is comes above A^n, Γ if neither the condition of (ω) nor (S) holds).

But (the reverse of) neither $>$ nor its transitive closure $>^*$ is guaranteed to be a derivation relation. This is because $>$ need not be well-founded, there may be an infinite sequence of sequents $\Gamma_1, \Gamma_2 \ldots$ such that $\Gamma_n > \Gamma_{n+1}$ for all n. On the alternative presentation of $>$ there may be an infinite upward path in the graph of sequents and horizontal lines. In a derivation, every upward path must end with an application of an *initial* sequent rule (from an empty sequent) such as (Z).

The definition of $>$ uses only computable relations between sequents. So for sequents Γ and Γ' *it is computable* whether or not $\Gamma > \Gamma'$.

Define the relation $>_\Gamma$ such that $\Gamma_1 >_\Gamma \Gamma_2$ iff $\Gamma >^* \Gamma_1$ and $\Gamma_1 >^* \Gamma_2$.[7] We now have the following facts:

1. $>_\Gamma$ is well-founded iff it is a tree

2. If $>_\Gamma$ is well-founded then it represents a derivation according to (ω), (S) and (Z) viewed as defining a derivation relation.

3. If we interpret sequents Γ as *disjunctions* of their (un-superscripted) members. Then $>_{A^0}$ is well-founded (a tree) iff A is derivable from ZFC together with the ω-rule.

I state these facts without proof, the proofs essentially apply the constructions of [5] which are expanded on in [1] (and also presented in a different context in [4]).

However some explanation is in order. Suppose that A is derivable from an axiomatisation of ZFC together with the ω-rule, then it has a derivation. The relation $>_{A^0}$ is a brute-force (infinitary) search for such a derivation. A sequent

---

[7]So $>_\Gamma$ is a restriction of the transtitive closure of $>$ to Γ.

such as A^n, Γ means that A is at the 'head' of a sequent and we have checked A against the first $n-1$ $\forall$-derivations. If the $n+1$th $\forall$-derivation d_n is a derivation that $\vdash_{\text{ZFC}} A$ then we have found what we are looking for and stop (with that particular path in the search). If d_n is neither a derivation that $\vdash_{\text{ZFC}} A$ nor $\forall x_{\in \omega} B \vdash_{\text{ZFC}} A$, by (S) we cycle A to the back of the sequent and increment its index to $n+1$ we then continue with the new head. If d_n is a derivation that $\forall x_{\in \omega} B \vdash_{\text{ZFC}} A$ then we branch our search to the infinitely many sequents $\Gamma, B[x/\mathbf{0}]^0, A^{n+1}$, again cycling A to the back and continuing to search for a derivation of it, but, in each new branch, we also search (back from the first derivation d_0) for a derivation of the relevant $B[x/\mathbf{0}]$. The fact that we are constantly cycling the sequent ensures that every formula in it gets checked against every derivation until one of them is derivable directly from ZFC. There is a derivation of A from the ZFC and the ω-rule if and only if this procedure terminates starting with the sequent A^0. That is, there is a derivation of A if and only if $>_{A^0}$ is well founded. Furthermore, it follows by induction on recursive ordinals that $>_{A^0}$ is well founded iff it is a tree (for any A).

Of course, except in the special case where $\vdash_{\text{ZFC}} A$, we could never represent A^0 as a tree (we cannot write down all the infinite branches). But that does not mean that we cannot describe it, we describe it as a tree. Thus we may adopt the following definition:

- A is true of ZFC when $>_{A^0}$ is a tree.

I claim that this can be asserted without explicit reference to quantification over ordinals or sets. To talk of derivations using the ω-rule requires induction up to an uncountable ordinal which presumes a predicate '...is uncountable' and '...is well founded' which themselves require a development of set theory (or second order logic). To define truth in terms of $>_{A^0}$ being well-founded is better, but requires quantification over sets to capture the property of well-foundedness. To say that $>_{A^0}$ is a tree however, we need only understand how to define which arbitrary relations are trees as was done in Section 5. I claim it does not require any prior development of set theory for anyone to understand the what it is to be a tree. I claim the definition is acceptable from a formalist point of view.

It might be thought that we can simplify the definition by saying that A is true of sets if it is derivable from ZFC plus the *recursive* ω-rule. Then the definition of derivability will require induction on the *recursive* ordinals rather than on ω_1. But how are we to characterise the recursive ordinals without quantifying over infinite sets when specifying they are well-founded? We can try to use our definition of trees noting that a recursive ordinal is what is common to isomorphic trees. But defining recursive ordinals as, e.g. equivalence classes of trees once again requires more infinite set theory. The notion of a tree seems prior to the notion of an (recursive) ordinal. We can define trees by an inductive definition, I claim, without any background set theory. Recursive ordinals are what isomorphic trees have in common, but we need set theory to say that.

We can expand on the proposed analysis of mathematical truth to one of mathematical consequence. Given a computable function f such that $f(n)$ is either a formula of $\mathcal{L}_{ZFC}$ or is 0, we can define a relation $>^f$ exactly as above but where (S) is modified to (S^f) as follows:

(S^f) $\quad A^n, \Gamma > \Gamma, A^{n+1}$ if $f(n)$ is *not* A, and the nth ∀-derivation is *neither* a derivation that $\vdash_{ZFC} A$ nor that $\forall x_{\in \omega} B \vdash_{ZFC} A$ for any B

We can then say that A is a *ZFC-consequence* of f (or the range of f, if we like) when $>^f_{\{A^0\}}$ is a tree. That is, A is a consequence of f if we allow the branches of a tree to terminate with the formulae of f as well as derivations from the axioms of ZFC.

7

We have been trying to offer a characterisation of set theoretic truth which incorporates the arithmetic truth definition of Section 1. The characterisation should not presume any more expressive power than was provided by the arithmetic truth definition which effectively allowed us only quantification over finite sets. Choosing ZFC as providing a base axiomatisation of the behaviour of sets, the definition we have arrived at is that A is true of ZFC when $>_{A^0}$ is a tree. Where 'tree' is an inductively defined property of binary relations.

Now that we have defined set-theoretic truth we can use it to verify that our truth definition is equivalent to our other proposed truth definition: that A should be true when it is derivable from ZFC plus the ω-rule. Indeed, the link between trees, recursive ordinals etc. can all be proved within ZFC (which is all true according to our truth definition).

The previous paragraph will have raised alarm. Firstly, if we can prove in ZFC that the definition of set theoretic truth in terms of trees is equivalent to the definition in terms of closure under the ω-rule, then does that not suggest that *both* presume some set theory if one of them does? Secondly, for any A, the relation $>_{A^0}$ is computable, and so can be expressed within ZFC, as can the definition of trees (or well-foundedness). So the assertion that $>_{A^0}$ is a tree can also be expressed within set theory. So we can produce, in $\mathcal{L}_{ZFC}$, a predicate of (Gödel numbers of) formulae A that appears to express that $>_{A^0}$ is a tree. Or, as ZFC derives is equivalent, a predicate T apparently expressing that x is the least set formulae closed under the axioms of ZFC, first order logic and the ω-rule. But then if the proposed truth definition is adopted, doesn't T become a truth predicate for $\mathcal{L}_{\text{ZFC}}$, which is impossible by Tarski's Theorem?

I shall address the second issue first. The reason T seems impossible is that, following Tarski's Theorem, we may construct a formula G such that $\vdash_{\text{ZFC}} G \leftrightarrow \exists x(Tx \wedge \mathbf{g} \notin x)$, where $\vdash_{\text{ZFC}} \exists! x Tx$ and $\mathbf{g}$ is the Gödel number of G for some suitable coding of the formulae of $\mathcal{L}_{\text{ZFC}}$. So by the truth definition $G \leftrightarrow \exists x(Tx \wedge \mathbf{g} \notin x)$ and $\exists! x Tx$ are true.

Let us write $t \in T$ as shorthand for $\exists x(Tx \wedge t \in x)$ and $t \notin T$ for $\neg \exists x(Tx \wedge t \in x)$ which is derivably equivalent in ZFC to $\exists x(Tx \wedge t \notin x)$. It follows that if the proposed truth definition is consistent (i.e. does not make all formulae true), and $\mathbf{n} \in T$ is true for all and only the Gödel numbers $\mathbf{n}$ of true formulae, we get a contradiction. If G is true then so is $\mathbf{g} \notin T$ which contradicts the assumptions about T. But if G is false then $\mathbf{g} \in T$, which again cannot be. The assumption we must drop is the assumption that $\mathbf{n} \in T$ is true for all and only the Gödel numbers $\mathbf{n}$ of true formulae, we may suppose (assuming the consistency of the truth definition) neither G nor $\mathbf{g} \in T$, nor their negations, is true.

But does not $\mathbf{g} \in T$ express that the formula G is the least set closed under the rules of ZFC and the ω-rule? No, it does not, in fact it does not express anything about sets, for there are no sets, this is a *formalist* theory of truth: $\mathbf{g} \in T$ are symbols that have meaning only in so far as we give them a truth condition. If the rules of ZFC have been well chosen, then, given a certain Gödel numbering, the truth conditions of $\mathbf{n} \in T$ are such that it is true iff $>_{A^0}$ is a tree, where $\mathbf{n}$ is the Gödel number of A. We should also expect it not to be the case that any A and its negation are both true. But, by reasoning similar to the above, ZFC will not be such that $\mathbf{g} \notin T$ is true if $>_{G^0}$ is not a tree.

The proposed semantics is therefore necessarily three valued, some formulae are true, some false, and some neither. If ZFC is a good choice, then G and $\neg G$ are examples of formulae with no truth value. This is not a problem, for both G and $\neg G$ make essential use of quantification over sets, which, the formalist proposal made here, has no background semantic structure giving it a definite truth value.

Returning to the first issue, the notion of a tree, I claim, has nothing to do with sets. There is however a way of simulating the trees within set theory. For example, to simulate the definition of Section 5 we must define, by induction on ordinals, what it is for an extension of a relation predicate (i.e. a set of ordered pairs) to be a tree. As mentioned above, if we have chosen the rules of ZFC well, then the truth conditions of certain assertions about trees will be the same as the truth conditions of their set theoretic simulations. The benefit of the simulation is that in the context of set theory, those same truth conditions may be expressed in a variety of different ways, and connections between other assertions (which may or may not be set theoretic simulations themselves) can be drawn. Moreover, the additional set theoretic language, gives us the ability to express additional propositions that were previously unexpressible (G is an example of such an expression, although perhaps not a particularly practical one). Just as with Arithmetic, the set theoretic language becomes an exercise in broadening our expressive power.

8

The truth theory for Arithmetic of Section 1 was formulated for a hybrid language $\mathcal{L}_A$ that had arithmetic and non-arithmetic parts. Thus far, the truth theory for ZFC has been presented in a language with no predicates or constants for real word objects and properties. What are the truth conditions of sentences about real world objects that make essential use of set theoretic vocabulary? If real numbers are given a set theoretic analysis, then we should hope to apply the analysis of the truth conditions of set theoretic statements assertions like:

Some rod's length in meters is transcendental.

Actually the problem is broader, for the proposed truth theory given for set theory is not compositional, but ought to be compositional for the natural language quantifiers. In the example above, and abusing notation a little, we expect 'some rod is F' to be true if, for some a, a is a rod and Fa. But the predicate '... is transcendental' in that sentence might require an (apparent) existential quantification over sets $\exists s A$ for which there is no term t such that $A[x/t]$ is true.[8] So the truth theory for the hybrid language must be compositional for the objectual parts but not for the set theoretic parts.

We could attempt to incorporate a compositional semantics, in terms of valuations, into the definition of $>$. But this has complications, in particular relating to the requirement that $>$ be computable: if A depends on some non-computable matters of fact (to be perfectly general I have no means of ruling out such a possibility), then $>_{A^0}$ may not be a tree, but that would be a poor reason to rule A as not true.

---

[8]By the proposed truth definition, $\exists s A$ can be true without there being any term t such that $A[x/t]$ is true, and $A \vee B$ can be true even though neither A nor B is true (although they cannot both be false). As already sketched, the truth definition is necessarily incomplete in that there are some sentences, e.g. G, which are neither true nor false. Thus, since $G \vee \neg G$ is true, we have an example of a failure of truth functional composition for disjunction. To see obtain an analogous result for existential quantification note that $G \vee \neg G$ is derivably equivalent in ZFC to $\exists s\big((s = 0 \to G) \wedge (s = 1 \to \neg G)\big)$.

There is, however, a more natural way of integrating the set theoretic and objectual languages which trades on the idea that set theory is a general abstract theory of relations and properties. A hybrid language will contain objectual predicates, $F, G \ldots$, functions $f, g \ldots$, variables $x, y \ldots$ and quantifiers etc. as well as the variables, quantifiers and $\in$ symbol of the set theoretic language. Given a sentence in the hybrid language A and convert it into a statement A', purely in the set theoretic language, about arbitrary set theoretic domains. Then we can say that if there is a purely objectual B, that is true according to the standard compositional semantics, such that $(B \to A)'$ is a set theoretic truth, then A is true. That is when reality says that some B is true, and set theory says that B implies A arbitrarily, then we can say that A is true. We can generalise on this:

- Given a suitable conversion function $'$. A hybrid sentence A is true when there is some computable property P of true objectual formula B, and A' is a ZFC-consequence of the formulae that are P.

What we want is a computable syntactic operation that converts any A into a suitable A'. To spell out such an operation fully is not hard, but beyond the scope of this paper. Instead I shall present an example. Let A be the following simple assertion:

$$\text{The set of people is nonempty.} \tag{A}$$

We have an objectual compositional semantics for an objectual language containing the predicates '...is a person', and a truth theory for a purely set theoretic language, but nothing for this hybrid assertion A. We first produce A'. We assume the language of ZFC contains a stock of n-ary predicate symbols (other than $\in$) for each n and let F be a unary predicate symbol. Now we set A' to be:

$$\exists x (\forall y (y \in x \leftrightarrow Fx) \wedge \exists z \in x) \tag{A'}$$

That is, for a uninterpreted predicate letter F, A' is the assertion that the set of F is non-empty. Now choose B to be:

$$\text{There is a person.} \tag{B}$$

And then set B':

$$\exists x Fx \tag{B'}$$

And now it is not hard to see that A' is a ZFC-consequence of B' and that B is fully interpreted by an objectual semantics (and is true).

The idea is that for a hybrid sentence A, A' is a similar sentence in the language of ZFC where the objectual predicates are replaced by uninterpreted proposition letters (or or more complex expressions of an implementation of $\mathcal{L}_A$). A is then true when A' is a ZFC-consequence of true objectual Bs, via their B'.[9]

9

In summary, I have presented here the beginnings of what I claim to be a formalist notion of set theoretic truth and its application in natural language. The truth conditions of arithmetic expressions have a straightforward compositional treatment, the truth conditions of sentences that use set theoretic terms do not. Some statements may be neither true nor false, nevertheless, their place in the underlying formal system (e.g. ZFC) gives them meaning. In short, the truths of set theory are the theorems of ZFC plus the infinitary ω-rule, although that is a set theoretic way of expressing it. I claim here that the truth definition can be expressed differently, without presuming any set theory.

References

[1] Torkel Franzen. Transfinite progressions: a second look at completeness. *Bull. Symbolic Logic*, 10(3):367–389, 09 2004.

[2] Michael Gabbay. A formalist philosophy of mathematics part I: Arithmetic. *Studia Logica: An International Journal for Symbolic Logic*, 96(2):219–238, 2010.

[3] D. Hilbert and P. Bernays. *Grundlagen der Mathematik*, volume I. Springer Verlag, Berlin, Heidelberg, New York, 1st edition, 1934. second edition with modifications and supplementations, 1968.

[4] Gerald E. Sacks. *Higher Recursion Theory*. Berlin: Springer-Verlag, 1990.

[5] J. R. Shoenfield. On a restricted omega-rule. *Bulletin L'Académie Polonaise des Science, Série des Sciences Mathématiques, Astronomiques et Physiques*, VII(7):405–407, 1959.

---

[9]More tricky are hybrid sentences such as:

$$\text{The length of rod } R \text{ in metres is } \pi \qquad (R)$$

$$\text{There are uncountably many people} \qquad (U)$$

The truth conditions (R) depends on an extension of the arithmetic language to express simple ratios of lengths. The truth conditions of (U) require a further discussion of the formalist notion of truth and the nature higher infinities. Both of these are beyond the scope of this paper.

A Non-Structuralist Alternative to Formalism

Danielle Macbeth
Department of Philosophy, Haverford College

Abstract

It is easy to see nineteenth- and twentieth-century debates in the philosophy of mathematics regarding formalism and structuralism as oscillating between, on the one hand, a mechanistic conception of mathematical *practice* in terms of the rule-governed manipulation of signs, and on the other, a conception focused instead on a *subject matter* for mathematics, in particular, on structures. I suggest that something more is at issue, something that is possible at all only in light of nineteenth-century developments in mathematical practice against the backdrop of Kantian philosophy. Both formalism and structuralism are responses to those developments, and both responses are, I argue, essentially Kantian. Although in the nineteenth century mathematical practice came to involve reasoning directly from concepts, and not the construction of concepts in pure intuition as Kant had held, nevertheless, the philosopher's understanding of what reasoning from concepts involves remained Kantian. The formalist argues directly: once we see that mathematical reasoning does not involve intuition, as Kant had held, we can see that such reasoning lacks all content, that it is purely formal (because, on the Kantian view, it is intuition that provides all content and all truth). The structuralist thought is further motivated by the realization that a system of axioms in mathematics originally formulated for one domain can be shown also to hold in other quite different domains. The structuralist thought is that we can treat the *uninterpreted* axioms as defining a structure. What remains quite unclear, however, is how uninterpreted axioms can serve to define anything. To resolve the difficulty, I suggest, we need a more radical, properly post-Kantian, post-model-theoretic alternative to standard structuralism.

Keywords: Frege, Model Theory, Nineteenth-Century Mathematics, Structuralism

---

This essay is a developed, and improved, version of the talk I gave at the conference "The Emergence of Structuralism and Formalism" in Prague in June 2016. I am grateful to the other participants for their very useful questions and discussion, and to the organizers for enabling our many fruitful exchanges over the days of the conference.

In a letter to Hilbert written in the fall of 1895, Frege outlines his understanding of the nature and value of symbols in mathematics; according to him, although "a mere mechanical operation with symbols is dangerous", mechanism is necessary in the practice of mathematics insofar as it "partly relieves the scientist from having to think" [4, p. 33]. Frege provides this analogy: "when a tree lives and grows it must be soft and succulent. But if what was succulent did not in time turn to wood, the tree could not reach a significant height" (Ibid.). Similarly, a subfield of mathematics that is living and growing ought not to be mechanized, ought not to involve the thoughtless manipulation of symbols according to rules. The reason is simple: in a still-developing subfield it is just not clear what exactly the signs are to mean and hence what the rules of their manipulation ought to be. Nevertheless, once the domain has been fully developed and adequately understood, systematizing it in a system of signs the manipulations of which are governed by antecedently stated rules can be very valuable, even essential to the ongoing development of the discipline. According to Frege, then, it is true that the science of mathematics has a subject matter but also true that systematization and mechanization are an integral part of its practice. Assuming Frege is right, it is not so surprising that we should find that philosophers, in their reflections on mathematical practice, have some tendency to focus on, or at least emphasize, one side of the dialectic at the expense of the other.

It is easy, I think, to see the nineteenth- and twentieth-century formalist – structuralist debate in such terms, as oscillating between, on the one hand, a mechanistic conception of mathematical practice, and on the other, a conception focused instead on a subject matter for mathematics, in particular on structures. In fact, I will suggest, something subtly different is at issue in this debate, something that is possible at all only in light of nineteenth-century developments in mathematical practice against the backdrop of Kantian philosophy. Early twentieth-century formalism in the philosophy of mathematics, it will be argued, was a very natural response to those nineteenth-century developments (against the backdrop of Kantian philosophy); and structuralism is, at least in part, a response to formalism, one that self-consciously aims to bring back into view a subject matter for mathematics.

But structuralism also responds directly to mathematical developments in the nineteenth century; it responds, in particular, to the discovery that axiomatizations developed for one mathematical purpose, in one mathematical context, can equally well be interpreted as serving that (or an analogous) purpose in a quite different mathematical context. What the structuralist takes this to show is that mathematics has, as it turns out, a quite distinctive sort of subject matter, namely, freestanding, or pure, structures. But the structuralist response to those developments is, we will see, not the only response possible. The structuralist, like the formalist, remains essentially Kantian. The more radical, properly post-Kantian response to be outlined here also deserves consideration.[1]

1 Developments in Nineteenth-Century Mathematics and the Formalist Response

According to Kant, mathematical practice involves an intuitive use of reason, the construction of concepts in pure intuition, and it is this activity of construction that is to explain the capacity of the mathematician to achieve synthetic a priori knowledge. The idea is not implausible. Consider, for example, the thought that the product of two sums of integer squares is itself a sum of integer squares. The mathematician does not merely reflect on the *concept* of a product of two sums of integer squares – as, Kant suggests, the philosopher might. Instead, the mathematician formulates the content of that concept in a collection of written marks, either in a diagram or in an expression of the symbolic language of arithmetic and algebra. The mathematician might, for example, exhibit the content of the concept of a product of two sums of integer squares thus: $(a^2 + b^2)(c^2 + d^2)$. This collection of signs is, in its way, an instance of the concept, though it is one that – unlike, say, '$(61^2 + 437^2)(555^2 + 79^2)$' – contains nothing that is not thought already in the concept. In just this way, on Kant's account, one goes outside one's concept to something with the form of an intuition. And now one can operate on the display as licensed by familiar rules.

$$(a^2 + b^2)(c^2 + d^2)$$
$$= a^2c^2 + a^2d^2 + b^2c^2 + b^2d^2$$
$$= a^2c^2 + b^2d^2 + a^2d^2 + b^2c^2$$
$$= a^2c^2 + 2acbd + b^2d^2 + a^2d^2 - 2adbc + b^2c^2$$
$$= (ac + bd)^2 + (ad - bc)^2$$

This last expression, we can see, has the form of a sum of integer squares; it is, schematically, an instance falling under the concept *sum of two integer squares*. And so we are done. Through our construction in what Kant thinks of as pure intuition, we have shown that the product of two sums of integer squares is itself a sum of integer squares; we have shown that being a product of two sums of integer squares, that *concept*, entails being a sum of integer squares, another concept. The two concepts are necessarily related, though not by logic alone. The concept of a product of two sums of integer squares does not *contain* the idea of being a sum of integer squares; nevertheless, as the construction shows, it does entail that latter

---

[1] I first discuss some of the themes to follow, especially those in parts II and III, in [7, §5.3, §6.2].

idea. The theorem we have proven is synthetic a priori, necessary, though not logically necessary.

According to Kant, mathematical practice generally is a practice of the construction of concepts in pure intuition as illustrated in our little example. But in fact, from its earliest beginnings mathematics has also involved reasoning directly from concepts. Consider, for example, the ancient Greek proof that there is no largest prime number. Because in ancient Greek mathematical practice there is no way to display the content of the concept of a prime number, *what it is* to be a prime number – as one *can* display what it is to be, say, a sum of integer squares, as above, or a circle, by drawing something roughly circular, and using it to draw inferences about the relative lengths of various radii of that circle[2] – one must reason directly from the relevant concepts. And so one does. First, we suppose that there are only finitely many prime numbers and that we have an ordered list of all of them. Now we consider the product of all those prime numbers and add one to that product. The resultant number is either prime or composite. If it is prime then obviously it is a prime larger than any on our list. But if it is composite then it must have a prime divisor larger than any on our list (because that composite number is a product of the primes on the list plus one). Either way, we have shown that there is a prime number larger than any on our original list. There can be no complete and finite list of all prime numbers as we had originally supposed; there is no largest prime. It is in just this way that one reasons directly from concepts in mathematics. And it is just this sort of reasoning directly from concepts – reasoning that can be *reported* in, say, spoken or written English but is not *displayed* in any specially devised system of written marks such as Euclidean diagrams or the symbolic language of arithmetic and algebra – that came in the nineteenth century to be the norm of reasoning in mathematical practice. As a direct result, it came to seem *clear* that Kant had been wrong about the practice of mathematics. Doing mathematics does not (constitutively) involve the construction of concepts in pure intuition but only what Kant thinks of as a discursive use of reason directly from concepts, by logic alone. And because (Kantian) logic abstracts from all relation to any object, and thereby from all content and all truth, it came naturally to seem that mathematics so practiced must be a purely *formal* discipline, the mere manipulation of signs according to rules of logic.

Other developments further reinforced the idea that mathematics had in the nineteenth century been *shown*, *pace* Kant, not to involve any appeal to intuition, and hence not to have any content at all insofar as it is, on Kant's view, only through intuitions that concepts, and thought more generally, have content. First, the very

---

[2]Euclidean diagrammatic reasoning is discussed at length in Chapter Two of [7].

idea of pure intuition was met with deep skepticism, for instance, by Bolzano and Herman Grassman. Already in 1810, in his *Contributions to a Better Founded Exposition of* Mathematics, Bolzano argues that the notion of a pure or a priori intuition is incoherent, that mathematical truths must be proven deductively from concepts. In 1817, Bolzano made good on his word: he proved the mean-value theorem deductively from definitions of the concepts it contains. Grassman, similarly, in his 1844 *Ausdehnungslehre*, recognizes only empty logical forms, forms that provide the basis for formal science, and empirical content providing the basis for real, empirical science. Pure intuition having been thereby banished, mathematics could be realized as a formal science. Struggles to deal with strange and counterintuitive but mathematically very powerful entities such as negative and complex numbers reinforced the idea insofar as attempts to find a place for such things in mathematics seemed also to push in the direction of a more formal conception of mathematics. Indeed, it was coming more and more to seem that all thought has two, and only two, distinct and separable parts, a formal part and a material part. We find, for example, George Boole arguing in his *Mathematical Analysis of Logic* (1847) that algebraic properties of multiplication and addition first formulated for natural numbers, can be shown to hold also for the intersection and union of classes, can be shown to hold even for the logical relations of conjunction and disjunction among propositions. The system of *signs* is determined solely by the rules governing their manipulations. It is a separate question, and one that admits of a variety of answers, how those signs are to be interpreted, what exactly they are to be taken to concern, whether numbers, or classes, or propositions.

Kant in the eighteenth century had recognized not only analytic judgments that are merely explicative, known by logic alone, and a posteriori judgments that are knowable only on the basis of perceptual experience, *but also* synthetic a priori judgments, which as necessary but not logically necessary are grounded in the forms of sensibility and understanding. By the end of the nineteenth century, in light of developments in the practice of mathematics over the course of that century, it had come to seem that in fact there are only analytic and a posteriori judgments, that all necessity is *logical* necessity, and hence, again, that mathematics is really nothing more than the manipulation of signs according to rules.

2 The Emergence of Structuralism

On the formalist view that is a direct consequence of rejecting constructions in pure intuition while leaving in place the rest of the Kantian architectonic, mathematics is nothing more than a play of symbols, strictly deductive reasoning that, as a matter of pure logic, is wholly formal. Mathematics, on this view, has no subject matter; lacking all relation to any intuition, it is devoid of all content and so all truth. It is merely the empty play of signs, very like a game constituted by the rules of play. But although various developments in nineteenth-century mathematical practice seemed to suggest just such a conception of mathematical practice insofar as they showed that the practice of mathematics does not make essential appeal to anything like Kantian constructions in pure intuition, there were other developments that seemed to suggest not that mathematics has no subject matter of its own, as the formalist holds, but that it has a new sort of subject matter. What those other developments seemed to show was that mathematics is not about objects – say, numbers (conceived as collections of units) and geometrical figures, as the ancient Greeks had thought – nor even about relations objects can stand in as expressed in equations, which was Descartes' view.[3] What those developments seemed to show is that mathematics is instead about a new sort of entity, a structure. We need to uncover the mathematical roots of this new, structuralist view of mathematical practice.

As indicated already in our very brief discussion of Boole's work in logic, it was becoming common in nineteenth-century mathematics and logic to distinguish between, on the one hand, a rule-governed system of signs, and on the other, an interpretation of that system. We find, for example, George Peacock distinguishing between what he calls arithmetical algebra, which is fully meaningful, about numbers, and symbolical, purely formal, algebra, which is wholly uninterpreted. Such a distinction is required, Peacock argues, because although symbolically it is perfectly correct to transform, say, the symbolic equation $x + b = a$ into the equation $x = a - b$, arithmetically this is intelligible only if a is larger than or equal to b. Arithmetical algebra, because it concerns numbers, is constrained by its subject matter, and in particular, by the fact (so it seemed to Peacock) that a larger number cannot be subtracted from a smaller number. Our concern now is with a different but related point, the fact that an axiomatization devised for one purpose can, given a different interpretation, be shown to hold for a different domain.

Among the significant advances in mathematics in the nineteenth century was the realization that the familiar axioms of elementary algebra, though devised first for ordinary numbers (the integers, the rationals, and the reals), can also be taken to be about quite different systems. Other systems of objects can be shown, that is, to be isomorphic to those numbers and as a result can serve as models for the axioms. We have, for example, these familiar axioms of elementary algebra.

1. $a + b = b + a$ Commutativity of addition

---

[3]For discussion of Descartes' view see [7] Chapter Three.

2. $(a + b) + c = a + (b + c)$ Associativity of addition

3. $a \times b = b \times a$ Commutativity of multiplication

4. $(a \times b) \times c = a \times (b \times c)$ Associativity of multiplication

5. $a \times (b + c) = (a \times b) + (a \times c)$ Distribution of multiplication over addition

As originally intended these axioms concern addition and multiplication as applied to ordinary numbers. But, again, they can be shown to hold also of other systems of numbers, for example, the integers modulo n. Because whatever theorems can be proved on the basis of the axioms must hold for any system that satisfies the axioms, it is natural to distinguish between the (uninterpreted) axioms and theorems, on the one hand, and an interpretation or model of them, on the other, to adopt, that is, a model-theoretic conception of language.

On the model-theoretic conception of language that distinguishes between, on the one hand, logical form, and on the other, content as given by an interpretation or model, a (first-order) language involves four essentially different sorts of signs that Hodges has characterized as follows.

> First, there are the *truth-functional connectives*. These have fixed logical meanings. Second there are the *individual variables*. These have no meaning. When they occur free, they mark the place where an object can be named; when bound they are part of the machinery of quantification.... The third symbols are the *non-logical constants*. These are the relation, function, and individual constant symbols. Different first-order languages have different stocks of these symbols. In themselves they don't refer to any particular relations, functions or individuals, but we can give them references by applying the language to a particular structure or situation. And fourth there are the *quantifier symbols*. These always mean 'for all individuals' and 'for some individuals'; but what counts as an individual depends on how we apply the language. To understand them, we have to supply a *domain of quantification*.
>
> The result is that a sentence of a first-order language isn't true or false outright. It only becomes true or false when we have interpreted the non-logical constants and the quantifiers [5, p. 144].

On this model-theoretic conception of language, a set of axioms such as those given above do not, independent of an interpretation or model, have any content or truth. It takes but one more step to achieve the insight of structuralism.

Consider, first, a different system, the familiar Peano axioms for the natural numbers:

1. $N0$

2. $(\forall x)[Nx \supset Ns(x)]$

3. $(\forall x)(\forall y)[(Nx \& Ny \& s(x) = s(y)) \supset x = y]$

4. $\sim(\exists x)(Nx \& s(x) = 0)$

5. $(\forall F)\{[F0 \& (\forall x)((Nx \& Fx) \supset Fs(x))] \supset (\forall x)(Nx \supset Fx)\}$.

Here again we can understand the language model-theoretically, in terms of a dichotomy of logical form and an interpretation or model that assigns a meaning to the non-logical constants, N, 0, and $s(\ )$, and stipulates a domain of quantification. The structuralist insight is that the set of (uninterpreted) axioms can itself be taken to define a pure structure, what we can think of as the natural number structure. That is, instead of assigning a meaning to the non-logical constants N, 0, and $s(\ )$, together with the stipulation of a domain of quantification, the axioms themselves are now to be taken to determine the contents of the non-logical constants N, 0, and $s(\ )$. And because this structure is wholly defined in purely logical terms, it is correctly thought of as a pure, that is, freestanding, structure. According to the structuralist, the subject matter of mathematics is just such pure or freestanding structures. On this view, then, mathematics *has* a subject matter despite the fact that it makes no appeal to pure intuition, despite the fact that it is, in its way, purely formal.

There can be no question that structures have been a central concern of mathematics since the nineteenth century. The group concept in abstract (modern) algebra is a paradigm. A group is any collection of elements C together with some operation * on those elements, closed under * (that is, if a and b are elements in C then $a * b$ is an element of C as well), that satisfies the following three constraints:

1. The operation * is associative: $(a * b) * c = a * (b * c)$

2. There is an identity element **i** in C such that $\mathbf{i} * a = a = a * \mathbf{i}$

3. Each element a in C has an inverse a' such that $a * a' = \mathbf{i} = a' * a$.

Notice that here there is no reference either to any particular elements or to any particular operation but only to certain properties of elements and operations. And the elements in the collection need not even be objects, though they can be. The elements can be, for example, a set of permutations of some thing or things, and the operation ∗ can be composition on permutations, that is, one permutation followed by another, which is associative. In that case, the identity element is the do-nothing permutation. And because for every permutation there is one in reverse, the third condition is satisfied as well. The permutations together with composition thus provide an instance of a group.

Another familiar example of a group is the integers under the operation of addition: any two integers can be added together to form another integer, the addition function is associative, zero serves as the identity element, and every integer has an inverse under addition, with zero as its own inverse. Similarly, the rationals (barring zero) form a group under multiplication. And these facts about the integers under addition and the rationals (barring zero) under multiplication, can explain significant facts about arithmetic such as that all linear equations of the form 'a + x = b' or 'ax = b' (a ≠ 0) are solvable. It is *because* these structures are groups that all the steps that are needed in order to solve such equations can be taken, and hence that those equations have solutions in every case.

Group theory has yielded up a vast range of important and unexpected theorems, and has revealed important connections between different parts of mathematics. Felix Klein, for example, famously used group theory to show why the quintic, not generally solvable in radicals, is solvable if appeal is made to a certain sort of complex function, and he did so by showing deep connections between the quintic, the theory of rotation groups, and the theory of complex functions. Group theory also has important applications in, for instance, physics, chemistry, and engineering. The notion of a group is exemplary of the sorts of structures that have formed the subject matter of mathematics since the nineteenth century.

3 An Alternative to Structuralism

As presented above, structuralism is very closely tied to the model-theoretic conception of language.[4] Structuralism is nonetheless not identical to model theory. Model theory draws an absolute distinction between logical form, on the one hand, and content provided by an interpretation or model, on the other. Independent of any interpretation or model sentences of the language conceived model theoretically have no content, no meaning, and no truth-value. But, it was suggested, the structuralist holds that it is the *uninterpreted* axioms that are taken to define a structure. But how, one might wonder, *can* the uninterpreted axioms of a language conceived

---

[4]Compare Steward Shapiro's [10] the aim of which is to show, among other things, that "the emergence of model theory and the emergence of structuralism are, in a sense, the same" (p. 144).

model theoretically serve to define a structure? The very real conceptual difficulties with the idea of an uninterpreted language defining a structure provide, I think, the underpinnings of the by-now very familiar debates within structuralism in the philosophy of mathematics concerning the reality of pure structures.

We saw that an important mathematical motivation for the model-theoretic conception of language was the fact that axioms originally formulated for one domain could be shown to hold also for other quite different domains. Provided that the axioms are satisfied in a given domain, the derived theorems will hold as well. Thus, as Hodges puts it, "for mathematical purposes, we need non-logical constants whose interpretation is fixed in a structure but not fixed in the language" [5, p. 146]. "We want to say, for example, that two structures have some property in common. One method is to write the property as a first-order sentence and say that it is true in both structures" [5, p. 145]. Hodges here seems to be assuming what is known as eliminative structuralism according to which there are no abstract structures defined by the uninterpreted axioms but only the various systems that exemplify the structure, that are models of the set of axioms in question. The *ante rem* structuralist wants something more, namely, the abstract structures themselves as defined by the uninterpreted axioms. For the *ante rem* structuralist, mathematical structures exist whether or not there are any systems of entities that exemplify them.[5] Indeed, it is precisely by *jettisoning* the content that is provided by an interpretation or model that the power of the pure form is revealed on this view. We read, for example, in Bourbaki's 1950 essay "The Architecture of Mathematics":

> From the axiomatic point of view, mathematics appears thus as a storehouse of abstract forms – the mathematical structures.... Of course, it cannot be denied that most of these forms had originally a very definite intuitive content; but, it is exactly by deliberately throwing out this content, that it has been possible to give these forms all the power they were capable of displaying and to prepare them for new interpretations and for the development of their full power.
>
> It is only in this sense of the word "form" that one can call the axiomatic method a "formalism". The unity which it gives to mathematics is not the armor of formal logic, the unity of a lifeless skeleton; it is the nutritive fluid of an organism at the height of its development, the supple and fertile research instrument to which all great mathematical thinkers since Gauss have contributed. ([2] quoted in [10, p. 177])

But, again, it is not at all clear that a model-theoretical conception of language *can*

---

[5]Shapiro's [10, p. 9] has a brief introduction to these varieties of structuralism.

support such a view. The relevant, Kantian, notion of logical form simply is not a kind of content. Rather, it contrasts absolutely with content. And we are back, so it seems, to a kind of formalism.

I claimed above that both formalism and structuralism are essentially Kantian. We are now in a position better to understand what this means. Although both the formalist and the structuralist reject any role for pure intuition in mathematics, both *keep* the Kantian conception of logic as purely formal, of quantification as involving reference to objects in a domain, and of content as provided by reference to objects. In short, the *very idea* of model theory – with its dichotomy of logical form and content provided by an interpretation or model that gives objects for the relevant forms to be about – is *deeply* Kantian. Thus, although the structuralist wants to provide a new subject matter for mathematics, that of pure or freestanding structures, it is not clear how this is possible. One inevitably falls back into formalism. We need another way.

Consider, again, this familiar law of elementary algebra: $a + b = b + a$. One way to understand this collection of signs is quantificationally, as implicitly quantifying over all the numbers in the domain. What the collection of signs says, so understood, is that for all pairs of numbers in the domain, the sum of the first and the second is equal to the sum of the second and the first. What we learned in the nineteenth century is that this identity holds also of quite different systems of objects. The question is how exactly to conceptualize what we have learned here. Model theory, we have seen, gives us one way. With Kant, we distinguish between the logical form, that is, the uninterpreted signs in the particular array $a + b = b + a$, and content as provided by an interpretation or model that assigns a meaning to the non-logical constants (here, the plus sign) and a domain of quantification to fix what objects the axiom is about. And now we can understand the claim that two structures both have the property that is of concern in this axiom. We say, following Hodges, that the sentence '$a + b = b + a$' is true in both structures, that the two structures are both models for the axiom. But again, this is not the only way to conceptualize this lesson of nineteenth-century mathematics.

On a quantificational reading, the sentence '$a + b = b + a$' is about objects, pairs of them, that the sum of the first and second is equal to the sum of the second and the first. Model theory extends this reading by introducing the idea that different systems of objects can provide an interpretation of such a form. A third and more radical reading of that same sentence has it that what the sentence is about is not *objects* at all but instead the *operation of addition*. On this reading, the sentence ascribes the higher-order property of being commutative directly to that operation. Objects do not come into it. What we think of as a quantifier is, on this new reading, a sign that, perhaps together with other signs, can be used to form expressions for just such higher-order properties as commutativity. The basic idea, which – as I have argued in [6] and in [7] – derives ultimately from Frege, can be illustrated already for a simple sentence such as 'Socrates is mortal'. As we naturally read this sentence it is about Socrates, the particular man, and says of that man that he is mortal. It ascribes the (first-order) property of being mortal to the man Socrates. But we can also read the sentence differently, as being about the (first-order) property *being mortal* and ascribing to it the (second-order) property of being a property of Socrates. Everything that is true of Socrates has a certain higher-order property in common, namely, that of being a property of Socrates, and what we are saying here, on our alternative reading, is that the property of being mortal has that higher-order property.

Suppose that we symbolize the sentence 'Socrates is mortal' as 'Ms'. On our first reading above, the sentence is analyzed into the first-level function $M\xi$ for argument s. But we have seen that we can also analyze the sentence differently, as involving the second-level function Φs, being a property of Socrates, for argument $M\xi$, being mortal. Similarly in the case of the sentence '$a + b = b + a$', although we *can* read it as about objects, more exactly, pairs of objects, that they have a certain property, we can learn also to read it differently, as about addition that it has a certain higher-level property, namely, the property of commutativity.

Perhaps it will be objected that there is still an oblique reference to objects on the second reading of '$a + b = b + a$' just as there is an oblique reference to Socrates on the second reading of 'Socrates is mortal' in the idea of the higher-level property *being a property of Socrates*. To see that there is no such reference in the arithmetical case, we need to think differently about generality than we have been taught to think in quantificational logic.

As classical logic teaches us to think of it, a sentence (of English) of the form 'all S is P' is a universally quantified conditional: for all x (in the domain of quantification), if x is S then x is P, that is, in standard symbolism: $(\forall x)(Sx \supset Px)$. On a Fregean conception of such a logical symbolism, the primitive signs only express senses independent of any context of use. We saw this already with the sentence 'Ms', which we can read as expressing a Fregean thought. On such a reading, it is only given a particular analysis into function and argument that we can say what the subsentential parts designate. Consider again the collection of signs: $(\forall x)(Sx \supset Px)$. Instead of reading it quantificationally – as saying that for all x (in the domain of quantification), if x is S then x is P – we can take it to express a Fregean thought, one that is by its nature subject to different analyses into function and argument. One natural analysis has it that the sentence is about the first-level conditional property *being P if S* that it has a certain higher-level property, namely, that of being *universally applicable*, of giving the value true no matter what object is provided as argument. On this reading, the argument is the first-level conditional property *being P if S*, and the function is the higher-level function *being universally applicable*, which takes first-level properties as arguments to give truth-values as values. But other analyses are possible as well. Suppose we take the two first-level properties *being S* and *being P* as the arguments. What we are left with in that case is a second-level relation, the relation of subordination that takes two first-level properties as arguments to yield a truth-value as value. Marking the argument places of this relation with uppercase Greek letters, we can designate this two-place second-level relation thus: $(\forall x)(\Phi x \supset \Psi x)$. The second-level, two-place relation that is designated by this collection of signs takes two first-level concepts, among them, *being S* and *being P*, as arguments to give a truth-value as value. It is a complex sign designating the relation of subordination.

Now we need to apply the basic idea to the sentence '$a + b = b + a$'. Here the argument place, on the analysis we are after, is marked by the two occurrences of the plus sign. Notice further that although there are two occurrences of this sign, the relevant function is a one-place function. That is, we have a case like that of 'Cato killed Cato' read as involving Cato as argument for the function *killing oneself*, which takes a single object as argument to give a truth-value as value.[6] So in our case, the argument is given by the two occurrences of the plus sign. The signs that remain when the signs for the argument are removed, replaced by something to mark the argument place – for instance, an asterisk: $a * b = b * a$ – together form a complex sign for the higher-order property *commutativity*. The sentence as a whole ascribes this higher-order property to the operation of addition.

Now we need to think about inference. And here again, I want to offer a way of thinking that is slightly different from the usual (Kantian) way in terms of a dichotomy of logical form and content. Whereas we generally think of the validity of inference in terms of logically valid forms, I want instead, following Ryle, Peirce, and others, to think of the validity of inference by appeal to the notion of a rule that is essentially general in applying also to indefinitely many other cases.[7] That is, we interpret the claim that an inference is valid in virtue of its form as saying that it is an instance of something more general, of a schema or rule that applies not only to the given case but also in other cases that are relevantly like it. For example, the inference from 'Felix is a cat' to 'Felix is a mammal' is good, if it is good, because one can infer generally from something's being a cat that it is a mammal. The inference is not good in virtue of being about Felix. Though one does infer something about Felix, the validity of the inference is explained not by

---

[6] Frege discusses this case in his 1879 logic [3, §9]
[7] See, for instance, Gilbert Ryle [9] and Charles Sanders Peirce [8, p. 131].

reference to Felix but by appeal to a rule, something to the effect that being a cat entails being a mammal. Similarly, the inference from the judgments that John is in the kitchen or the garden and that John is not in the garden to the judgment that John is in the kitchen is good, if it is good, not in virtue of being about John and his whereabouts. Though one does infer something about John and his whereabouts, the validity of the inference is explained not by reference to John but by appeal to a rule, that known as disjunctive syllogism. Whereas in our Felix example the rule concerns valid inferences that rely on the meaning of 'cat', in the John example the rule concerns valid inferences that rely on the meanings of 'or' and 'not'. And just the same is true in all other cases of actual (valid) inference, whether material or formal: any actual (valid) inference is an instance of a rule that applies also in other cases.

On the conception of the goodness of inference just outlined, an inference is valid not in virtue of form *as contrasted with content*, but instead in virtue of a conception of form that is itself a kind of content. That is why we could take both formally valid inferences, such as our inference about John and his whereabouts, *and* materially valid inferences, such as our inference about Felix, as *inferences* properly speaking just as they stand. Consider now the fact that the axioms of (say) elementary algebra, on the current reading, ascribe higher-level properties to the operations of addition and multiplication. Now we derive various theorems on the basis of those axioms. Those theorems ascribe *other* higher-level properties to those operations. The reasoning, in other words, has the form: given that (as set out in the axioms) this and that are properties of addition and multiplication, those operations, addition and multiplication, must also have the properties so-and-so and such-and-such (as stated in the theorems). Much as we inferred that Felix has the property of being a mammal given that Felix has the property of being a cat, so here we infer that addition and multiplication have various properties given that they have the properties ascribed in the axioms. And now we can see that the inferences from the axioms to the theorems are not good in virtue of the fact that they are about addition and multiplication – any more than the inference about Felix is good in virtue of being about Felix. They are good because having those particular properties that are ascribed in the axioms entails having other properties as well. That is just what the derivations of the theorems from the axioms show. It follows directly that any other function or operation having the same properties as are ascribed in the axioms (commutativity, associativity, and so on) will also have the properties that are ascribed in the theorems that are derived from the axioms. The insight that the axioms and theorems can be applied in other cases does not, then, show that the model-theoretic conception of language is correct. That insight can as easily be conceived as a reflection of a fundamental feature of inference, the fact that any particular inference is an instance or application of a general rule that can be applied also in other cases.

As we have read them here, the axioms of elementary algebra are contentful truths about the operations of addition and multiplication from which other truths follow. But again, the inferences we make from the axioms to the theorems are not good in virtue of being about addition and multiplication. They are good in virtue of something general that applies also to other cases. *Any* operations that satisfy the axioms, that have the higher-level properties ascribed in the axioms to addition and multiplication, will have also the higher-level properties ascribed in the theorems. And so it is with our little example above to show that the product of two sums of integer squares is itself a sum of integer squares. As Avigad notes, "the proof uses only the commutativity and associativity of addition and multiplication, the distributivity of multiplication over addition and subtraction, and the fact that subtraction is an inverse to addition; hence it shows that the theorem is true much more generally in any *commutative ring*" [1, p. 115]. The little chain of reasoning rehearsed above in the symbolic language of arithmetic and algebra is valid, so we know that it is an instance of something more general, of a rule that can be applied also in other cases provided that the requisite properties hold in those cases. This chain of reasoning does not show that we are not, in this case, reasoning about products of sums and sums of squares. What it shows is that the reasoning can also be applied in other cases as well. Again, *any* actual inference is an instance of something more general, something that can be applied in other cases as well. It is just this that explains the fact that, as Avigad notes, our theorem about integer squares holds in any commutative ring.

4 Conclusion

In the nineteenth century the practice of mathematics underwent a profound transformation, the second such transformation in its long history, the first having been begun with the introduction of Descartes' analytic geometry in 1637. Indeed, as Stein famously remarked, this second transformation was "so profound that it is not too much to call it a second birth of the subject" [11, p.238]. And as I have argued in [7], with this second transformation reason, pure reason, is fully realized as a power of knowing. The practice of deductive reasoning from concepts that has since the nineteenth century been the norm in mathematical practice is the work of pure reason alone. It is also, as Frege saw, ampliative despite being deductive. Frege's task was to understand how that can be, and it was in coming to understand such ampliative deductive proof that Frege made a radical break with the

Kantian conception of logic as purely formal, empty of all content. On Frege's new, essentially post-Kantian conception, even logic is a substantive science with its own subject matter. Logical language is not, then, to be conceived model theoretically – though not because it is always already interpreted as is sometimes claimed. What is needed, Frege sees, is not the distinction of logical form and content as provided by a interpretation or model, but instead the distinction of *Sinn*, sense expressed, and *Bedeutung* or designation. It is just this understanding, together with a new conception of the validity of inference, that enables us to come to a substantially new understanding of the developments in nineteenth-century mathematics that had seemed to lead at once to model theory and structuralism.

Is this new account really just itself another form or species of structuralism? In a way it clearly is – and should be. From the perspective provided by actual mathematical practice, it seems manifest that much of contemporary mathematics concerns structures, in particular, pure or freestanding structures, structures defined solely by logically articulated properties. On this point I wholly agree with the *ante rem* structuralist. What is distinctive about the proposal outlined here is that it clarifies how this *can* be. Having jettisoned model theory, replaced it with a properly Fregean conception of language as I have articulated it, first in [6] and then further still in [7], we are able to show just how a system of axioms can articulate a pure structure by appeal to various higher-order properties. Although one might first discover, say, the group concept through considering cases, by a kind of Aristotelian naïve abstraction, the concept in its mathematical use is instead to be given a top-down characterization in terms of the higher-order properties that are constitutive of a group. The fact that the instances have those properties is to be seen as secondary. And such top-down characterizations are ubiquitous in contemporary mathematical practice. What I have tried to show here is that it is Frege – who, of course, was himself a nineteenth-century mathematician – who helps us to understand how such pure, freestanding structures can in this way be given in an axiomatization that provides a top-down characterization in terms of higher-order properties, forming thereby the subject matter of the discipline of mathematics as it has been developed since the nineteenth century.

References

[1] Jeremy Avigad. Mathematical method and proof. *Synthese*, 153(1):105–159, 2006.

[2] Nicholas Bourbaki. The architecture of mathematics. *The American Mathematical Monthly*, 57(4):221–232, 1950.

[3] Gottlob Frege. *Begriffsschrift: Eine Der Arithmetische Nachgebildete Formelsprache des Reinen Denkens*. L. Nebert, 1879.

[4] Gottlob Frege. *Philosophical and mathematical correspondence*. University of Chicago Press, Chicago, 1980. Edited by Gottfried Gabriel et. al.; abridged from the German edition by Brian McGuinness and translated by Hans Kaal.

[5] Wilfrid Hodges. Truth in a structure. *Proceedings of the Aristotelian Society*, 86:135–151, 1985.

[6] Danielle Macbeth. *Frege's Logic*. Harvard University Press, Cambridge, 2005.

[7] Danielle Macbeth. *Realizing Reason: A Narrative of Truth and Knowing*. Oxford University Press, Oxford, 2014.

[8] Charles S. Peirce. *Reasoning and the Logic of Things: The Cambridge Conferences Lectures of 1898*. Harvard University Press, Cambridge, 1992. Edited by Kenneth L. Ketner.

[9] Gilbert Ryle. "If," "So," and "Because". In Max Black, editor, *Philosophical Analysis: A Collection of Essays*. Cornell University Press, Ithaca, 1950.

[10] Stewart Shapiro. *Philosophy of Mathematics: Structure and Ontology*. Oxford University Press, Oxford, 1997.

[11] Howard Stein. Logos, logic, and logistiké: some philosophical remarks on nineteenth-century transformation of mathematics. In *History and philosophy of modern mathematics*, volume 11 of *Minnesota studies in the philosophy of science*, pages 238–259. University of Minnesota Press, Minneapolis, 1988.

Are *Ante Rem* Structuralists Descriptivist or Revisionist Metaphysicians? How We Speak About Numbers

Prokop Sousedík
Catholic Theological Faculty, Charles University

David Svoboda
Catholic Theological Faculty, Charles University

Abstract

Ante rem structuralists endorse descriptive metaphysics. Accordingly, Shapiro takes what mathematicians say seriously (*faithfulness* constraint) and at the same time does not wish to revise the results they reach (*minimalism* constraint). When arithmetic is viewed from this perspective, one reaches the conclusion that a number is a relational object and therefore structuralism is justified. However, difficulties arise once the limits of common arithmetic are crossed. Then it turns out that numbers are also operated in ways that contradict structuralism. That can lead one either to doubt structuralism as a whole, or to reject descriptivism. We prefer the latter alternative, whereby we reject descriptivism only partially. It remains valid in the context of well-established mathematical practice, let us say arithmetic; it poses difficulties in spheres that as yet lack clear contours, for instance when mathematicians say $2_{real}=2_{nat}$. Although they understand statements of such type, we think that the way they express them is misleading and often confused. And we think that in such circumstances philosophers have the right to take part in creating more exact means of expression. Obviously, this proposal weakens Shapiro's minimalism constraint.

Keywords: ante rem structuralism, cross-structural identity, descriptivism, S. Shapiro, M. Resnik

In the history of philosophy of mathematics, two substantially different ways of thinking about the ontological status of numbers can be distinguished. We will call the representatives of one *descriptivist metaphysicians* and the representatives of the other *revisionist metaphysicians*.[1] Tentatively it is possible to say that descriptivist metaphysicians consistently start from mathematical practice and confront the results they reach with it. For revisionist metaphysicians, on the other hand, mathematical practice is not binding and as a result their conclusions often contradict the spirit of mathematic practice in different ways.

Although ante rem structuralists do not make use of the distinction inspired by Strawson, their way of speaking shows that they would reject revisionism and prefer descriptivism. For example, according to Resnik *mathematical realists are not committed to claims about mathematical objects beyond those they hold by virtue of endorsing the claims of mathematics* [11, p. 92]. Shapiro speaks in a similar way when he places two requirements on a philosopher of mathematics. According to one we are committed by the way in which mathematicians speak (*faithfulness constraint*); according to the other we ought not to associate any properties with mathematical objects other than those that mathematicians associate with them (*minimalism constraint*) [13, p. 110]

But the emphasis on descriptivism brings serious difficulties. Critics point out first of all that there are contexts in which mathematicians speak in ways which contradict the conclusions reached by structuralists. Or put differently, there are contexts which defy the relational character of mathematical objects. Such criticism can mean two things: Either structuralism is incorrect as such, or the descriptivist point of departure is incorrect. In our paper we want to present a comprehensive overview of the arguments employed by the opponents of structuralism and indicate a solution that would be immune to the above objections.

In the first part of our paper we will explain in what exactly the descriptivist point of departure consists and at the same time we will show how it brings philosophers of mathematics to structuralism. In the second and third part we will discuss the arguments against structuralism and at the same time we will show how they can be addressed.

1 Descriptivism – a path to structuralism

In order to delineate the descriptivist approach in a concise way, it will be useful to focus on the late Wittgenstein, whose *Philosophical Investigations* contain a short

---

[1] The distinction between descriptive and revisionist metaphysics was introduced by P. F. Strawson, cf. Individuals: An Essay on Descriptive Metaphysics. We leave the original meaning of this distinction aside and adopt merely its terminology. Investigating the relationship between descriptivism in mathematics (as formulated perhaps by Shapiro) and descriptivism as it is understood by Strawson would certainly be interesting, but is beyond the scope of this paper.

commentary on Augustine's reflections on time. We will regard this as a paradigm of how to proceed in our reflections on the ontological status of numbers [15].

Augustine, as Wittgenstein has noted, regards the problem of time as simple and difficult at the same time. *Something that we know when no one asks us, but no longer know when we are supposed to give an account of it* [15, §89-90]. The situation is similar when it comes to clarifying other philosophical concepts, including the concept of number. When no one is asking us, we all know well what numbers are, we operate with them, as with time, every day; but when someone asks us, we are as puzzled as we are about the nature of time.

Wittgenstein then shows how questions such as "What is time?" (and also "What are numbers?") can be dealt with. He first points out that we need to remind ourselves of something. But not of the world of ideas, as the Platonic-minded Augustine would probably have recommended, but of *the kind of statement that we make about phenomena* [15, §90]. When we speak about time we must recall to mind *the different statements that are made about the duration, past present or future, of events* [15, §90]. Similarly, when we speak about numbers we need to remind ourselves of statements in which numbers play a key part.

After this introductory reflection we now focus on the conception of number which ought to result from reminding ourselves of statements in which numbers play a key part. We will start from common equations of the type $7 + 5 = 12$. Platonic-minded philosophers (viz. *ante rem structuralists* as well) mostly hold that mathematicians do not verify the veracity of such propositions experientially, it is given *a priori*. They further point out that such propositions express identity and that numerals are therefore singular terms signifying individual objects.[2]

These two observations together result in the rejection of all forms of aposteriorism, represented by Aristotle and later by J. S. Mill. According to the adherents of this (revisionist) approach, mathematics, despite all appearances, does ultimately depend on experience and numerals are universal terms. We reject such aposteriorism and endorse Platonism, which is in better accord with the descriptivist point of departure, or more precisely with Shapiro's faithfulness constraint.

But the practice of mathematicians suggests that endorsing Platonism involves difficulties. These become patent when one recalls the characteristic of Platonism offered by M. Resnik. According to him, a traditional Platonist is *someone who holds that ordinary physical objects and numbers are "on a par." Numbers are the same kinds of thing – objects – as beach balls, only there are more numbers than beach balls and numbers are abstract and eternal* [9, p. 162].

---

[2]This is how Aristotle interprets Plato's conception; we think that in contemporary philosophy this approach was initiated by G. Frege.

So it seems that, according to a traditional Platonist, physical objects (beach balls) and numbers differ only in having a different ontological status: the former perish, the latter do not. If this ontological difference is left aside, then there should be no essential difference between the number seven and a particular beach ball. As descriptivists, we ought to confirm that in that a particular beach ball is spoken about in more or less the same way as the number seven. That, however, is not the case! A beach ball can be found in various contexts once it is the property of some retailer, later it becomes the property of my daughter Clare. So first it is true that the ball belongs to a retailer, later it is not true because it belongs to Clare. But when such ownership (and other) relationships change, the ball does not, it is still the same thing. So the ball is not part of relationships to things of the same kind necessarily, but merely contingently so. Therefore these relationships cannot belong to the ball's essence. On the contrary, the ball's essential properties are those that are not determined by context (by relationships to external things), but by the ball's internal composition.

Let us now take a look at what the situation is like in the case of some number, perhaps the number seven. Like the beach ball, the number seven is found within relationships to things of the same kind, but these things can only be other numbers. So it is, let us say, true that $7 > 5$, 7 follows 6, etc. But while in the case of the beach ball it is possible to think that it ceases to belong to a certain retailer, in the case of the number seven it is inconceivable that it would lose its relationship to the number six. So unlike the ball it is impossible to think of a situation in which the relationships of the number seven to other numbers would change. Therefore, the relationships between individual numbers are necessarily associated with these numbers, whereas the relationships between individual physical objects are contingent. Of course, that is not the only difference. We already know that properties capturing its internal composition are predicated of the beach ball, when we for example say that it is green or made of rubber. But only relational properties are predicated of the number seven by arithmeticians. So a ball has an internal composition, which ultimately determines its essence, but the number seven does not. So the essence of seven can only be given by relationships to other numbers.

Such considerations lead us to conclude that a number and a beach ball are not, as a traditional Platonist might think, on a par and traditional Platonism must therefore be rejected. For according to Resnik *In mathematics ... we do not have objects with an 'internal' composition arranged in structures, we have only structures. The objects of mathematics ... are structureless points or positions in structures* [10, p. 530]. But according to traditional Platonists mathematical objects do have an internal composition and cannot be regarded as *structureless points*.

S. Shapiro speaks in a similar way, since according to him the essence of a natural number is determined exclusively by *its relations to other natural numbers. The subject matter of arithmetic is a single abstract structure, the pattern common to any infinite collection of objects that has a successor relation with a unique initial object and satisfies the ... induction principle* [12, p. 72].

So the way we speak about numbers shows that their identity (unlike the identity of physical objects) is determined exclusively by their relational context and they cannot be taken out from it. So mathematical objects have no internal composition and are not on a par with physical objects.

2 The problems of structuralism

As we have stated above, the starting point of structuralism is descriptivism, i.e., it does not want to revise mathematical practice, merely to reflect on it. However, its critics point out that the structuralist conception ultimately contradicts descriptivism. For when we recall the statements we make about numbers, we find that some of them contradict their relational essence. We divide the critical objections against structuralism into three groups:

(a) Numbers are frequently associated with accidental properties (*8 is a number which is significant for Czech history*).

(b) Numbers are associated with philosophical properties (*8 is an abstract object*).

(c) Mathematicians identify objects belonging to different structures, i.e., they make statements of the type $2_{nat} = 2_{real}$.[3]

In this part of our exposition (2) we will address the problems associated with the accidental and philosophical properties of numbers, in the subsequent section (3) we will focus on the problem of statements of the type $2_{nat} = 2_{real}$.

Now we will concentrate on statements in which numbers are associated with accidental properties. These can be empirical properties, e.g. 8 has the property *to be important for Czech history*, or non-empirical properties, e.g. 2 has the property *to be the number of the solutions of the equation* $x^2 = 4$. If we accept the descriptivist point of departure, we should either include these properties in the characteristic of numbers, or explain which properties belong to the characteristic of numbers and which do not. The former appears unacceptable. When we ask what is eight, i.e., what is the essence of that number, then in searching for the answer we can probably leave aside that it is important for Czech history. In the second case the situation is

---

[3] We have not here space enough to deal with the problem of automorphism and that is why we leave it aside. There are some papers on the problem, see e.g. [14].

similar and relationship to the number of solutions of the equation $x^2 = 4$ also does not seem to belong to the essence of 2. If we therefore want to know what numbers are, we must consider only their essential properties and leave accidental properties aside.

But how are we to distinguish between essential and accidental properties of numbers? How do we know that the property *to be greater than seven* is essential for eight, while the property *to be important for Czech history* is accidental? Perhaps we could suggest that essential properties are necessary for numbers, while accidental properties are contingent. But this criterion can be successfully applied only in the case of empirical properties and it fails in the case of non-empirical properties. Two necessarily has the property *to be the number of solutions of the equation* $x^2 = 4$, and yet the property is not essential for the number two. A different criterion must therefore be found.

But how are we to go about that? In some way, we should probably know beforehand what belongs to the essence of a number and what does not. Being structuralists, we know that the essence of a natural number is given by its relationships to other natural numbers. Neither a relationship to history, nor a relationship to the number of solutions of some equation meets this requirement, which is why they are accidental properties. But the problem of this solution is obvious: we are simply going round in a circle. In order to distinguish between essential and accidental properties, we must know beforehand what these properties are.

So how are we to avoid being trapped in a vicious circle? We think that it is necessary to insist that we know the essence of numbers, albeit merely implicitly, not explicitly (i.e., we do not know beforehand that it is defined relationally). For, to return to Augustine, a competent speaker knows what numbers are (or what time is) when no one is asking him (he knows it implicitly, i.e., he uses numerals correctly), but does not know when asked (he does not know it explicitly, i.e., he does not know what is characteristic of such correct usage). Now if we want to find statements expressing the essence of numbers or time, we must find a competent speaker. In the case of the concept of time each of us is a competent speaker, but with the concept of number the situation is more complicated. Although everyone uses numbers frequently, familiarity with this usage is not acquired, so to speak, with mother's milk (as in the case of the concept of time), but at school while being instructed in arithmetic. So it appears that with respect to the essence of number the competent speakers are arithmeticians when they are speaking as arithmeticians. Therefore, arithmeticians know best what numbers are, but we must not ask them about it, we must merely observe how they speak as arithmeticians. Therefore, when we as philosophers ask what numbers are, we must recall statements made by arithmeticians. Concerning statements of the group (a), which were to set the structuralist definition of number in doubt, it must be stated that an arithmetician speaking as an arithmetician would never seriously utter them. Therefore they are not essential properties, but accidental ones.

Let us now consider objection (b), which points out that numbers are said to have philosophical properties, e.g. *to be abstract, to be of a relational character*, etc. These statements also contradict the structuralist definition of number. The fact that 7 is abstract is essential for it in a certain sense, but *to be abstract* is not a relational property. We think that the problem can be solved by distinguishing between mathematical and philosophical properties. A mathematical property of 8 is e.g. that it is greater than 7, its philosophical property is that it is perhaps an abstract object. It is principally possible to distinguish between mathematical and philosophical properties in a similar way as in the previous case. We recall statements made by arithmeticians as arithmeticians and contrast them with statements made by philosophers of mathematics.

The difficulty of our solution again consists in that it is problematic to distinguish between arithmetical and philosophical statements. For philosophical statements can be encountered in arithmetical treatises, mathematicians also utter them, but – unlike accidental properties – they cannot identify them. Therefore, in order to distinguish between mathematical and philosophical properties with certainty, we would need to find an ideal mathematician (or an ideal philosopher) who would admix nothing philosophical (or mathematical) in their discourse.

Obviously, an ideal mathematician or philosopher is not to be found and we must therefore simplify the solution to our problem in some way. Our proposal consists in selecting a well-established part of mathematics, concerning which there are no doubts. A good arithmetician safely operates with numbers, infallibly determines their properties and, apart from the introductory parts of his treatise, does not infect his practice with problematic philosophical statements.

Of course, it could be objected that a precise criterion, with the help of which it would be possible to distinguish between statements of philosophy of mathematics and statements of mathematics itself, is still missing. Perhaps even a mathematician would argue that statements of the type *7 is an abstract object* do belong to his mathematical practice. As philosophers we would perhaps argue that statements of such type belong in the sphere of philosophy and support our argument by saying that we are competent speakers in philosophy, while he is not. Of course, such a discussion can fail when both sides fortify themselves in their positions. But in many cases a philosopher can influence a mathematician in a certain way and clarify his speech. Let us say that the mathematician acknowledges that the statements *a number is an abstract object* belongs not to mathematics, but to philosophy, and subsequently "bothers" his readers with statements of this kind only in the introductory chapter of his book.

Such reasoning can lead us to the conclusion that a very fruitful relationship could exist between mathematicians and philosophers. A philosopher should "supervise" statements made by mathematicians and point out to them that something foreign has penetrated their field. Thereby he would of course not only make their speech more perspicuous; he would at the same time facilitate the development of their discipline.[4] At first glance this view of the relationship between philosophy of mathematics and mathematics seems to be quite natural. But from the structuralist point of view it is associated with a difficulty. For if we accept that a philosopher influences mathematical speech in a certain way, we thereby cast a certain amount of doubt on structuralism's descriptivist point of departure. For it seems that in some circumstances a philosopher can revise mathematical practice. But that would explicitly contradict Shapiro's first requirement, according to which mathematical speech is trustworthy (*faithfulness constraint*).

We think that the *faithfulness constraint* must be weakened to some extent. In our view a philosopher is not bound by mathematical speech in all circumstances. We will attempt to justify this view in the following paragraph.

3 Cross-structural identity problem

Another problem casting a shadow of doubt on the descriptivist starting point of structuralism concerns the identity (or relationship) of objects from different mathematical structures. The doubts are raised by statements of the type $2_{nat} = 2_{real}$. In order to understand in what the problem of such statements consists, one must recall that in mathematics *we do not have objects with an 'internal' composition arranged in structures, we have only structures* [10, p. 530]. But if the objects of mathematics have no internal structure, then they have no identity or features outside a structure. That means that statements of the type $2_{nat} = 2_{real}$ are strictly speaking nonsensical, because there can be no identity conditions relating to different things. Therefore statements of such type have no truth value.

But the problem is that mathematics works with a number of mutually distinct structures: there is the structure of natural numbers, of rational numbers, of real numbers, etc. Often, these different structures are mutually compared or embedded

---

[4] That such a relationship is possible is evinced by the history of physics. That discipline had been infiltrated by the metaphysical substantial conception of space. As a result, physicians spoke as if space (and also time) were absolute, i.e., quite independent of the world around us. E. Mach pointed out that physics had thereby been infiltrated by something metaphysical and that due to this confusion physics could not develop properly. It is generally known that this observation was an important step towards the theory of relativity. Cf. e.g. [8, Ch. 5].

in one another. On such occasions mathematicians frequently identify an object from one structure with an object from another structure and make statements of the type $2_{nat} = 2_{real}$[5]. That, however, ought not to be possible, because mathematical objects have *no identity or features outside a structure*. Therefore, a truth value can be associated with the statement $7 + 5 = 12$, because the statements 7+5 and 12 signify objects belonging to the same structure, but not with the statement $2_{nat} = 2_{real}$, because the expressions 2_{nat} and 2_{real} signify objects belonging to two different structures. So although according to the structuralist conception it is not possible to identify objects belonging to one structure with objects belonging to another structure, mathematicians in fact do make such cross-structural identity statements.

At first sight the above argument seems to be a "checkmate". For if mathematicians really make statements of the type $2_{nat} = 2_{real}$, then in accordance with descriptivism they ought not to be doubted and it should be admitted that mathematical objects do have an identity or features outside a structure. But, as we have shown above, a purely relational object cannot have an identity beyond the respective structure. If, on the other hand, we point out to mathematicians that statements of such type have no meaning or truth value, and therefore they ought not to make them, we cast doubt on our descriptivist starting point, and thereby on the results we have reached. What path should we take in these circumstances? We think that a structuralist has no other option than to weaken descriptivism – as in the preceding considerations (cf. 2) – and admit a certain influence of philosophy on the way mathematicians speak. More precisely, it is necessary to weaken Shapiro's *faithfulness constraint* and keep his *minimalism constraint* valid. The following exposition is divided into two conceptual parts. In the first one we will focus in greater detail on statements of the type $2_{nat} = 2_{real}$; in the second one we will attempt as philosophers to come to terms with the fact that mathematicians do make such statements.

At first sight it seems that statements of the type $2_{nat} = 2_{real}$ are not meaningless, since it should in principle always be possible to determine whether an object is identical with some other object. So arbitrary objects can be substituted for a and b in the identity form $a = b$ and the resulting statements will be true or false. That

---

[5] The fact that mathematical practice ultimately leads to statements of this kind is pointed out by F. MacBride [7, p. 568], quoting G. Kreisel [6]. It allegedly occurs when mathematicians want to grasp the properties of one domain of objects d by embedding it in a more extensive domain d^*. Mathematicians are very often obliged to establish ... embeddings between the objects described by different mathematics theories in order to establish results that would otherwise be unobtainable ... But embedding one domain in another assumes that some objects in these two domains are identical.

is how Frege thought of the matter, when he held that $a = b$ must have a meaning even when Caesar is substituted for a and 2 for b (*Caesar = 2*). We think that the solution to this famous *Caesar problem* presents the key to a proper understanding of the cross-structural identity problem.

Frege held that numbers are individual objects belonging to one universe which *consist of all that there is, and it is fixed* [3, p. 71]. But if there is one universe, then for any identity statement, there is a determinate fact of the matter as to whether it is true or false. We must therefore have a criterion, whose application will allow us to determine the truth value of statements of the type *Caesar = 2*, and also of the type $2_{nat} = 2_{real}$. Since there is one universe, from the semantic point of view the two statements are of the same kind as, e.g., the statement $7 + 5 = 12$. But how does a mathematician determine whether the statement *Caesar = 2* is true? The statement $7 + 5 = 12$ is verified by a common arithmetical procedure, but we lack a verification method to determine whether *Caesar = 2* or $2_{nat} = 2_{real}$ are true and can only be guided by intuition.

In order to be able to verify such statements, according to Frege, we must present a definition of number. He attempted to present it in two substantially different, albeit mutually interconnected ways. First he searched for a criterion of the identity of numbers. In the spirit of descriptivism he reminded himself of so-called *number statements*, i.e., statements of the type *the number of moons of Jupiter = 4*. By reflecting on number statements he concluded that the criterion of identity ought to be equinumerosity. But statements of the type *Caesar = 2* resist the criterion. He therefore reconsidered the definition and concluded that numbers are to be identified with a certain kind of extension (the class of equinumerous classes). That makes it possible to determine the truth value of the statement *Caesar = 2*, for if we can answer the question what numbers are, then we understand that two is not Caesar.

From our point of view it is important to realize that in a certain respect Frege proceeded very much like structuralists do. He and they in fact start from the descriptivist starting point, i.e., remind themselves of statements of a certain kind. Before Frege's or the structuralists' reflections we knew what numbers are only implicitly, i.e., when no one was asking us; after them we have an explicit answer, i.e., we can react when someone asks us. Frege's and the structuralists' answer to the question "What are numbers?" is therefore philosophical in Augustine's sense. If we therefore say that a number is a class of a certain kind or a purely relational object, we speak not as mathematicians, but as philosophers; we have not associated a mathematical property with numbers, but a philosophical one.

There the similarity between Frege's logicism and structuralism ends. In solving the question "What are numbers?" Frege started from number statements, while structuralists start from statements of pure arithmetic. The logicist answer to the question "What are numbers?" was not, unlike the structuralist one, purely philosophical (i.e., it was not good only for solving the Caesar problem); it also inspired theorists searching for the foundations of mathematics. Once it has been established that in fact every area of mathematics can be reduced to a theory of sets, set-theoretical hierarchy came to be regarded as an ontology valid for all of mathematics. That will make it possible to work with only one kind of objects in mathematics and the discipline will thereby attain remarkable unity. As a result, Frege's philosophical answer to the question what numbers are was to transform the entire mathematical practice so far in a significant way.

But, as P. Benacerraf observed [1], the project of logicism gave rise to two mutually interconnected difficulties. First: identifying numbers with sets gives rise to questions which have as yet not been touched upon by the reflections of traditional arithmeticians. For if numbers are sets then we can, for example, ask whether $2 \in 4$ or $2 \notin 4$. So Frege's reflections on the foundations of arithmetic have enriched traditional arithmetic. For we have discovered that totally new (set-theoretical) properties are to be associated with numbers. And while from the descriptivist point of view (or more specifically, from the point of view of Shapiro's *minimalism constraint*) this conclusion is problematic, a logicist need not necessarily endorse descriptivism. He would certainly point out that numbers do have set-theoretical properties and he has discovered them. Second: There are several mutually distinct identifications of numbers with sets: according to von Neumann $2 = \{\emptyset, \{\emptyset\}\}$ holds, according to Zermelo $2 = \{\{\emptyset\}\}$ holds. And these identifications have mutually distinct consequences: according to the former $2 \in 4$ holds, while according to the latter $2 \notin 4$ holds. But if these identifications have different consequences, then we are obliged to decide which one to endorse, for we must necessarily know which enrichment of arithmetic is the correct one. But such a decision cannot be made and therefore numbers cannot be identified with set-theoretical objects.

But the impact of Benacerraf's reflections is broader. Not only is it unacceptable to identify numbers with set-theoretical objects; it is unacceptable to identify numbers with any objects at all, for if a number is identified with some object, it thereby acquires its properties. But that results in the same problematic consequences as in the case of identifying numbers with sets. Identifying numbers with some objects would result in enriching arithmetic and again we are unable to determine whether such enrichment would be correct or not. Benacerraf therefore believed that numbers are not objects, which is why they cannot be identified with any objects whatsoever. Statements of the type $2 = \{\emptyset, \{\emptyset\}\}$ or *Caesar* = 2 are nonsensical, because they contradict the logical syntax of language: while *Caesar* or $2 = \{\{\emptyset\}\}$ are singular terms, the number 2 is not.[6]

One could object that Benacerraf does not account for the fact that two kinds of objects could exist: objects whose essence is defined with an internal structure and objects whose essence is defined exclusively in relational terms. But according to structuralists mathematics operates exclusively with the latter type of objects, which is why in mathematics *we do not have objects with an 'internal' composition* .... These objects have no *identity outside a structure* [10, p. 530]. But if numbers are objects having no identity outside a structure, then there is no fact of the matter based on which a truth value could be assigned to propositions of the type *Caesar* = 2. The statement *Caesar* = 2 is therefore nonsensical not because two is not an object (as Benacerraf held), but because the number two and Caesar belong to two mutually incompatible categories. This conclusion is in good accord with descriptivism (faithfulness constraint): mathematicians speaking as mathematicians would never utter statements of the type *Caesar* = 2. So a consequence of these reflections is the incommensurability of natural numbers with objects of the external world. But if there is an unsurpassable abyss between numbers and objects of the external world, then Frege's assumption of one universe must be set aside and a plurality of universes must be admitted.[7]

The existence of a plurality of mutually incommensurable universes can also be confirmed by applying the descriptivist point of departure. As mathematicians do not interpolate statements of the importance of the number eight for Czech history into arithmetic (cf. §2), so they do not infect arithmetic with statements of the identity of Caesar and the number two. Textbooks of arithmetic contain neither *8 is important for Czech history* nor *Caesar* = 2. The same is true of statements of the type $2 = \{\emptyset, \{\emptyset\}\}$.

After these reflections we can now focus on statements of the type $2_{nat} = 2_{real}$. These are, apparently, of the same kind as *Caesar* = 2 or $2 = \{\emptyset, \{\emptyset\}\}$, for natural numbers belong in a different general category than real numbers and statements of this kind are again nonsensical. For if we identified 2_{nat} with 2_{real}, then we would add the properties of 2_{real} to the properties of 2_{nat} and thereby we would again enrich arithmetic in an inadmissible way. So Benacerraf's argument can be successfully applied not only to the identification of natural numbers with set-theoretical objects, but also to the identification of natural numbers with real numbers.

---

[6] In our context we would define the logical syntax of language in a way similar to R. Carnap in his famous paper *The Elimination of Metaphysics Through Logical Analysis of Language*.

[7] That could ultimately be admitted even by Benacerraf, according to whom identity statements make sense only in the contexts where there exist possible individuating conditions ... [Q]uestions of identity contain the presupposition that the entities inquired about both belong to some general category. [1, §III.A]

This conclusion should be confirmed (as above) by rigorously applying descriptivism. But, as we have already pointed out, this is where an unpleasant conflict arises. While mathematicians do not say $Caesar = 2$ or $2 = \{\emptyset, \{\emptyset\}\}$, they *sometimes find it convenient, and even compelling, to identify the positions of different structures* [12, p. 81] So they do make statements of the type $2_{nat} = 2_{real}$.

According to Shapiro, Resnik attempts to come to terms with this problem in the following way: "*if the structures are kept separate, there is no fact of the matter concerning whether their places are identical or distinct from each other* [13, p. 125]. So a truth value cannot be associated with statements of the type $2_{nat} = 2_{real}$ without further consideration. But if the respective structures are not kept separate, but are rather both considered in the context of an encompassing theory, then a truth value can be associated with the statements.

The first shortcoming of Resnik's solution consists in that it abandons the original realist point of departure. If there is no fact of the matter, based on which it would be possible to determine whether statements of the type $2_{nat} = 2_{real}$ are true or not, then such a fact of the matter cannot begin to exist when a theory of a higher kind is created. A philosopher, who believes that he creates facts of the matter by his own mental activity, is not a realist, but an idealist. A second shortcoming is that it ultimately again gets into conflict with the descriptivist point of departure, because it tells mathematicians that to make statements of the type $2_{nat} = 2_{real}$, they must introduce a new kind of universe. In this way Resnik attempts to refine mathematical expression, whereby he comes into conflict with Shapiro's *faithfulness constraint*.

Shapiro invests greater effort in the problem of identity of 2_{nat} and 2_{real}. His first attempt again starts from the assumption that there is no fact of the matter concerning whether the numbers 2_{nat} and 2_{real} are identical or distinct from each other. If we nonetheless make statements of the type $2_{nat} = 2_{real}$, we must interpret the symbol "=" differently than in identities of the type $7 + 5 = 12$. In common equations there is a matter of fact which makes them true, but in statements of the type $2_{nat} = 2_{real}$ there is none. That is why allegedly *cross-identifications like these are matters of decision, based on convenience* [12, p. 81].

Shapiro further reflects on an objection which has been raised by Kastin in an unpublished text. According to this objection, Shapiro's solution does not accord with structuralist realism. For if mathematical objects exist, as objects not of our making, then identities between them cannot be stipulated. Among existing objects, identities are discovered, not made. So the allowed stipulations are inconsistent with the realism in ontology (similar objection can be found also in [4, p. 924] and [5, p.192]. The objection appears to be of a similar kind as the first objection against Resnik. If statements of the type $2_{nat} = 2_{real}$ had a truth value, the original realism would collapse into idealism. A fact of the matter which makes it possible to assign a truth value to $2_{nat} = 2_{real}$ arises by our decision to identify 2_{nat} with 2_{real}. We could further object to Shapiro, as we have to Resnik, that his proposal does not accord with the minimalism constraint. It is obvious that the symbol "=" is used differently in statements of the type $7 + 5 = 12$ than in statements of the type $2_{nat} = 2_{real}$. If that is the case, then we should again point out this ambiguity to mathematicians and ask them to make their terminology more precise.

Shapiro is aware especially of the first objection, so he amends his conception (Shapiro 2006). Kastin's objection leads him to claim that statements of the type $2_{nat} = 2_{real}$ are false. We think that this solution is not acceptable. If a statement of the type $2_{nat} = 2_{real}$ is to have a truth value, then nothing should prevent statements of the type Ceasar $= 2$ or $2 = \{\emptyset, \{\emptyset\}\}$ from having a truth value. So for every identity statement a fact of the matter should exist, based on which we would determine the truth value of the relevant identity. But if all identities can be meaningfully determined, then all objects are mutually comparable and can therefore be meaningfully included in one universe. But if all objects can really be included in one universe, then the properties of natural numbers are not given only by their mutual relationships, but also by relationships to other objects which are not natural numbers. So it would be a property of, let us say, the number two not only that it follows one, that it is not five, etc., but also that it necessarily is not Caesar or that it is not the real two. But arithmeticians do not speak of such properties; they were only discovered by philosophers. So when we claim that $2_{nat} \neq 2_{real}$, then we have as philosophers discovered a new mathematical truth. That, however, is an offence against descriptivism, or more specifically against the minimalism constraint.

However, let us assume for the sake of argument that Shapiro's solution is correct and that it therefore holds that $2_{nat} \neq 2_{real}$. That brings along a difficulty, namely that mathematicians accept $2_{nat} = 2_{real}$ and reject $2_{nat} \neq 2_{real}$. Shapiro therefore claims that in fact mathematical expressions are ambiguous. According to him the expression 2_{nat} in $2_{nat} = 2_{real}$ does not signify a place in the structure of natural numbers, but a place in a substructure of real numbers. If precisely this meaning is assigned to the term 2_{nat}, the equation becomes true. But accepting this solution again results in a contradiction of the *faithfulness constraint*. We should point out to mathematicians that expressions of the type 2_{nat} are ambiguous and that the ambiguity ought to be removed.

The difficulties associated with Resnik's and Shapiro's solution have prompted our own proposal, which already brings us to the second conceptual part of this paragraph. We agree with Shapiro's and Resnik's proposal that no fact of the matter making it possible to assign a truth value to statements of the type $2_{nat} = 2_{real}$ exists. But how are we to grapple with the fact that mathematicians do make statements of such a kind?

We think that one must first of all bear in mind that mathematics develops and so does the mathematical way of speaking. It is therefore possible to distinguish between a well-established mathematical practice and a practice which is, at least in its way of speaking, as yet unstable and often confused. Arithmetic is an example of the former group of mathematical practices and infinitesimal calculus in 17th and 18th century is an example of the latter. A philosopher ought not to approach these two groups in the same way. He ought to regard the spheres which are well established as faithful and approach the as yet unstable areas with some caution.

Of course, we need to ask how we are to distinguish between the two areas. The first criterion everyone will want to use is pragmatic. We will regard the traditional mathematical fields as faithful, while the newly emerging ones will necessarily raise some doubts. Although this criterion is useful, on its own it is unreliable. A new field is problematic and doubtful merely because it is new, while an old one is reliable merely because it is old. A philosopher therefore ought to look for another criterion. We think it is necessary to ask whether the mathematical way of speaking is consistent. That is what G. Berkeley[8] did in *The Analyst* and that is ultimately what we are doing here. Berkeley noticed that mathematicians are inconsistent in wielding the concept of increment,[9] we are pointing out that they are inconsistent in wielding the concept of number. When they operate with numbers in arithmetic, they conceive of them as of purely relational objects having no f; when, on the other hand, they make statements of the type $2_{nat} = 2_{real}$, they conceive of numbers as of objects with an internal composition.

With this finding we can conclude our reflections (as Berkeley did), or we can attempt to understand (as Resnik and Shapiro have) what exactly is meant by that type of statements and propose a way in which mathematicians ought to speak in order to be consistent.

We will take the latter path and allow ourselves to be inspired by Frege's *Grundlagen*. There the author reflects on sentences answering the question *How many?*, i.e., so-called number statements. There are essentially two equivalent answers to this question: When we ask *How many persons are there in the room?*, we answer either *The persons in the room are three* or *The number of persons in the room is three*. In everyday communication the two statements function in essentially the same way, but from the philosophical point of view the first answer is misleading, while the second one is not. *The persons in the room are three* can lead us to conclude that it is an identity statement, identifying the aggregate of persons in the

---

[8] Berkeley famously described infinitesimals as the ghosts of departed quantities. See [2].
[9] Cf. [2, p.25].

room with the number three. But such a conclusion is hardly acceptable, since it results in identifying numbers with objects of our world, i.e., with objects belonging to an entirely different general category.[10] Therefore, the second answer is to be preferred and number statement is to be identified with the statement *The number of persons in the room is three*. In this statement we are not identifying numbers with real aggregates; by means of language, with which we speak about the world around us, we are unambiguously describing an abstract object which does not belong to our world. The expression *The number of persons in the room* is a description referring to the same (abstract) object as the numeral *three*. So the general form of number statements which is not misleading is *the number of F(x)=n*. Numbers are not associated or identified with objects of the external world, but with (sortal) concepts which are used to describe the objects of the external world.

Having recalled Frege's procedure, we can now consider statements of the type $2_{nat} = 2_{real}$. We shall assume that the statements *Caesar = 2* and $2_{nat} = 2_{real}$ are misleading for the same reason: they identify objects belonging to different general categories. However, the statement *Caesar = 2* differs from the statement *The persons in the room are three* in that a competent speaker would never seriously utter it. If we did hear it, we would certainly think that it was uttered by someone who was as yet learning the language, e.g. a child or a foreigner. If we take a charitable attitude to the speaker, we will wonder what he might have meant. It could certainly occur to us that it is in fact a number statement, i.e., an answer to the question *How many Caesars are there?*. Obviously, the question and the answer (*Caesars are two*) are both misleading. For Caesar is only one! But at the same time it is necessary to say that bearers of the name may be more numerous. So our incompetent speaker is apparently answering the question *How many bearers of the name "Caesar" are there?* and should adequately say *The number of bearers of the name "Caesar" is* 2.

We will take a similar attitude to the mathematician who says $2_{nat} = 2_{real}$. Again it is a statement which a competent speaker ought not to say. For it – like *Caesar = 2* – identifies objects belonging to different general categories. But in this case the speaker is neither a child, not a foreigner. Yet the case is analogical! Above we have distinguished between mathematical practice which is well-established and one which is not. With the statement $2_{nat} = 2_{real}$ our speaker has apparently left the former and entered the latter. He has become, so to speak, a mathematical child or a foreigner. His way of speaking is inaccurate and can mislead the listener to conclude that it is possible to identify objects belonging to two different general

---

[10] Frege rather thought that the statement has a subject-predicate structure (the subject are the persons in the room, the predicate is the number three).

categories. But let us again be charitable and ask what exactly the speaker had in mind!

If the statements $Caesar = 2$ and $2_{nat} = 2_{real}$ are misleading in the same way, then $2_{nat} = 2_{real}$ must be interpreted in the same way as we have interpreted the term *Caesar* above. Therefore, if we have shown that the term *Caesar* essentially functions as a description of the abstract object 2, then $2_{nat} = 2_{real}$ must also be a description of the same object. Statements of the type $2_{nat} = 2_{real}$ are therefore number statements having the form *the number of* $F(x) = n$. But the expressions in this description are not used to describe the world around us; they are terms we use to speak about the structure of real numbers. The sortal concept $F(x)$ is thus a part of the language we use to speak about real numbers. Further investigation would require determining the character of the sortal concept $F(x)$.

These reflections show that when embedding one structure in another, mathematicians are in fact not identifying mutually corresponding objects, but applying objects of a poorer structure (natural numbers) to a richer structure (real numbers). And from the semantic point of view this application does not differ from applying natural numbers to the world around us.

4 Conclusion

Above we have indicated that the problem of cross-structural identity can be approached in alternative ways. From our point of view it is significant that all suffer of a common shortcoming. Each, albeit in different ways, alerts mathematicians to the imprecision of their speech and in fact asks them to make their way of speaking more precise. This follows up on the doubts which entered our considerations when we were reflecting on the dangers of the merging of mathematical and philosophical discourse in §2. There we also pointed out to mathematicians that their speech is infected by something foreign and can therefore be confused.

We think that such considerations cannot cast doubt on *ante rem* structuralism itself; we understand them as an incentive towards a re-evaluation of descriptivism. We believe that it is necessary to insist on Shapiro's minimalism constraint, while somewhat revising his faithfulness constraint. We think that only a well-established mathematical practice is faithful. That implies that some mathematical expressions are not part of this practice and a philosopher therefore has the right to doubt them and propose revisions. In this way a philosopher can participate in the development of mathematical investigation.

References

[1] P Benacerraf. What numbers could not be. *Philosophical Review*, 74:44–73, 1965. reprinted in Benacerraf and Putnam, *Philosophy of Mathematics* , pp. 403-420.

[2] G Berkeley. *The Analyst: a Discourse addressed to an Infidel Mathematician*. London, 1734.

[3] J Heijenoort. Logic as calculus and logic as language. *Synthese*, 17:324–330, 1967.

[4] G Hellman. Review of Shapiro. *Journal of Symbolic Logic*, 64:923–6, 1999.

[5] G Hellman. Three varieties of mathematical structuralism. *Philosophia Mathematica*, 9(3):184–211, 2001.

[6] G Kreisel. Informal rigour and completeness proofs. In I. Lakatos, editor, *Problems in the Philosophy of Mathematics*, pages 138–186. Amsterdam, North-Holland, 1967.

[7] F MacBride. Structuralism reconsidered. In S. Shapiro, editor, *The Oxford Handbook of Philosophy of Mathematics and Logic*. Oxford, OUP, 2005.

[8] E Mach. *Space and Geometry in the Light of Physiological, Psychological and Physical Inquiry*. Open Court, La Salle, 1960. Trans. by T. J. McCormack, original work published 1901-1903.

[9] M Resnik. *Frege and the Philosophy of Mathematics*. Cornell University Press, New York, 1980.

[10] M Resnik. Mathematics as a science of patterns: Ontology and reference. *Nous*, 15, 1981.

[11] M Resnik. *Mathematics as aăScience of Patterns*. Cornell University Press, New York, 1997.

[12] S Shapiro. *Philosophy of Mathematics, Structure and Ontology*. Oxford University Press, Oxford, 1997.

[13] S Shapiro. Structure and identity. In F. MacBride, editor, *Identity and Modality*, pages 109–145. Oxford University Press, Oxford, 2006.

[14] S Shapiro. Identity, indiscernibility, and ante rem structuralism: The tale of i and $-i$. *Philosophia Mathematica*, 15(3):1–24, 2007.

[15] L Wittgenstein. *Philosophical Investigations*. Blackwell, Oxford, 1953.

What Sort of Mathematical Structuralism is Category Theory

Josef Menšík
Masaryk University, Brno, Czech Republic
mensik@mail.muni.cz

Abstract

Although category theory is highly relevant for mathematical structuralism, its relevance does not consist in providing any new general philosophical approach to the philosophy of mathematics. On the contrary, category theory can be naturally interpreted as an exercise in either implicit, formalist, model, universals or modal structuralism – all of the five most general versions of mathematical structuralism recognized.

Keywords: Mathematical structuralism, Category theory.

1 Introduction

Within the philosophy of mathematics, the last thirty years have witnessed a rise to prominence of mathematical structuralism. Having established itself as the leading approach, mathematical structuralism still has to resolve some of its own issues. One of these is the relation of category theory to mathematical structuralism. Although I for one regard category theory as highly relevant for mathematical structuralism, in this paper I want to concentrate on where category theory does not bring anything fundamentally new. To be more precise, after presenting five main versions of mathematical structuralism I will explain why I do not think category theory to constitute any separate new one. The main claim of the paper is that category theory actually represents an example of all of the five general types to mathematical structuralism recognized.

Several classifications of different versions of mathematical structuralism were introduced in the literature (cf. e.g. Carter, 2007, Cole, 2010, Hellman, 2001, 2005, Horsten, 2012, Landry, 2016, Reck, 2003, Reck and Price, 2000, Shapiro, 2011, 2014). These classifications usually list around three to four different positions each, while being partly overlapping. Putting the various classifications together, I have been able to identify five main versions of mathematical structuralism which I am going to consider presently. In the rest of the paper I will explain why category theory, though recognized in some of the lists as a separate version of structuralism, in fact forms a specific example of all of these five general approaches to mathematical structuralism.

2 Implicit structuralism

The main idea of mathematical structuralism is quite straightforward: mathematics deals with structures (or patterns), rather than isolated objects, and it is the structural relations as perceived by means of these structures that count in mathematics, rather than other, non-structural features of the situation in hand.

Present time mathematicians are, at least implicitly, well aware of this character of mathematics. This is due to the prevalent use of axiomatic method in contemporary mathematics. The figures most often credited with introducing structuralist views into the philosophy of mathematics are the two major proponents of axiomatic method: Richard Dedekind and David Hilbert (cf. Landry, 2016, Mac Lane, 1996, p. 176., Shapiro, 2014).

A detailed evaluation of Dedekind's structuralism was provided by Reck (2003). Reck has explained how Dedekind addressed the system of natural numbers in an axiomatic fashion. Dedekind speaks about "freeing the elements [of the natural numbers system] from every other content" and "entirely neglecting the special character of the elements." While calling his axioms "conditions," he explains that because "relations and laws ... are derived entirely from the conditions" they "are therefore always the same in all [such systems]" (Reck 2003, p. 376).

Dedekind thus used axiomatic approach to clarify structural features of the system of natural numbers. He also maintained that any "internal" or "non-structural" features of its elements are irrelevant from the algebraic point of view. Moreover, he explained why the same (structurally derived) propositions were true in any system specified by the same axioms, these systems being thus in this sense isomorphic (cf. also Weaver, 1998, pp. 257, 260).

Hilbert is well known for setting up the whole program of axiomatization in mathematics, after himself showing the way by axiomatizing Euclidean geometry. Shapiro (2005, p. 64) states that in 1891 Hilbert is quoted as saying that in properly axiomatized geometry "one must always be able to say, instead of 'points, straight lines, and planes', 'tables, chairs, and beer mugs'." Some 10 years later he tried to explain the same idea in a letter to Frege:

> ... it is surely obvious that every theory is only a scaffolding or schema of concepts together with their necessary relations to one another, and that the basic elements can be thought of in any way one likes. If in speaking of my points, I think of some system of things, e.g., the system love, law, chimneysweep ... and then assume all my axioms as relations between these things, then my propositions, e.g., Pythagoras' theorem, are also valid for these things ... [A]ny theory can always be applied to infinitely many systems of basic elements. One only needs to apply a reversible one-one transformation and lay it down that the axioms shall be correspondingly the same for the transformed things. (Hilbert's to Frege, December 29, 1899, quoted from Shapiro, 2005, p. 66)

Hilbert thus explained that the names of mathematical concepts are irrelevant and that it is only the structural relations among the concepts (or things) that matter for a mathematical theory.[1]

To sum up, if a mathematical system is given by means of implicit definitions, such as in the case of an axiomatically given system, it only is the structural features of the system as resulting from these implicit, relational definitions what matter for the mathematical theory in question. Let me exploit the fact that this position is based on implicit definitions of concepts, and that it forms an implicit background to all the other structuralist positions, and call it *implicit structuralism*.

Reck (2003, pp. 370–74) calls the same position "methodological structuralism," and explains that it regards mathematicians as concerned with studying of:

> ... systems of objects, of both mathematical and physical natures ... which satisfy certain conditions: the defining axioms ... More precisely, they study such systems *as* satisfying these conditions, (Reck, 2003, p. 370)

While introducing three other forms of structuralism, he finds them all grounded in implicit structuralism (his methodological structuralism). According to Reck all three "can be seen to be guided by, or at least compatible with, methodological structuralism". He also remarks that otherwise these other positions "differ significantly" regarding the "abstraction" which is present behind them (Reck, 2003, p. 374).

Correspondingly, Carter (2007, p. 123, 2005, p. 306) explains that it is only structure as "described by a collection of axioms" or "described by a theory" that captures

---

[1] Apart from Shapiro (2005), another discussion of Hilbert's structuralism is available in Landry (2009).

all the possible notions of structure in mathematics. Any further elaborations on the concept of structure are accordingly bound to lead to dissent.

Lets us now move past the agreement among structuralists, as captured by implicit structuralism, and study the differences between various specific structuralist positions.

3 Formalist structuralism

Mathematics as captured by a formal language is concerned with formally written sentences and proofs. It is the patterns of terms construction and inference patterns which constitute the object of formal mathematics. The formalist position in the philosophy of mathematics, which explicitly recognizes the former, is thus intrinsically structural. *Formalist structuralism* is concerned with structures of mathematics as they are recorded in a formal language.

I have borrowed the term "formalist structuralism" from Reck and Price (2000), though, I am taking into account only the third variant of theirs (cf. Reck and Price 2000, pp. 347–8). I am in agreement with their observation that this approach is "seldom recognized explicitly as [a] structuralist position by contemporary philosophers" (Reck and Price 2000, p. 374). Indeed, among all the lists of different structuralisms quoted above, formalist structuralism only appears once, on the very list of Reck and Price.

4 Model structuralism

The main idea of *model structuralism* is to understand mathematics as a study of structured systems built up from other pre-supplied mathematical entities. The idea behind this is that if we want to pursue the valid-in-all-models way of thinking, we have to specify what counts as a model. The most common example of model structuralism is the set theory structuralism, where all the structured systems are represented by sets equipped by some extra structure, "certain relations or operations" Hale (1996, p. 124). For the set-model structuralist, "[s]tructures are understood as models (sets as domains with distinguished relations and possibly individuals)" (Hellman, 1996, p. 102).

This approach has strong tradition in mathematics itself because it mirrors the development of mathematics in the 20th century. After set theory was first axiomatized in early 20th century, it became a standard background theory for the rest of mathematics. Although the set represents perhaps the simplest structure imaginable – that of a collection of unordered or otherwise unrelated items – it was quickly realized that one may construct relations, functions and operations out of sets. A typical example of structured system thus became an "algebraic structure" built on a set such as the group: a set equipped with operations of composition and inversion, and with one distinguished element (the unit), all satisfying further appropriate conditions. Other examples of "structures build on sets" are ordered sets, graphs, vector spaces, topological spaces, monoids, rings and many others.

When some 20th century mathematicians went to look for "the foundations of mathematics", they, quite naturally, found them in the set theory. The whole endeavour culminated with the Bourbaki group programme to ground all of mathematics on set theory.[2]

In the usual set theory axiomatizations the elements of sets are other sets. This means that the elements of structures defined over sets are sets. Convenient as it may be from some other points of view, it introduces a potential problem for mathematical structuralism. Sets used as model elements carry with them their own internal structure which thus becomes unwanted part of the model structures. Together with the structural features, which any model shares with all other models of the same structure, any model has also its distinguished features which are irrelevant from the structuralist point of view. These are the features introduced by means of the specific sets used in the construction of the model.

This problem was first pointed out by Benacerraf (1965) in connection with natural numbers. Since numbers modelled in set theory are sets, we may ask set theoretical questions like "is number 2 an element of number 3" or "how many elements does number 3 have". Such questions certainly have nothing to do with algebra, but they are, nevertheless, perfectly legitimate questions about a set theory model. The answers to these questions will critically depend on the particular sets we have used in the model.

This problem concerns model structuralism generally, not only the set theoretical structuralism. If we want to use other pre-supplied mathematical entities to construct the structure-exemplifying models by means of them, we should be aware that the features of the entities used are transferred with them into our models.

The moral is that two concrete models of a structure which were constructed from different entities are, in principle, distinguishable by some structurally irrelevant features in which they differ. Although we would like to call two models exemplifying the same structure as structurally isomorphic, we cannot do so in the context of the background theory which supplied the model entities.

A hypothetical way to address this problem would be to try to abstract over all models when speaking about structures, and consider as structural only those

---

[2]For more on "Nicholas Bourbaki" cf. e.g. Mac Lane, 1996, p. 179.

features which are shared by all the models. But this already moves us away from the model structuralism concerned with concrete structures in the direction of a further form of mathematical structuralism we are about to examine presently.

5 Universals structuralism

While the first position considered above, implicit structuralism, is perhaps tacitly present behind all structuralist approaches, the second and third positions are rarely regarded as properly structuralist within the present discourse. Let us now proceed to an approach actually widely explicitly discussed by contemporary mathematical structuralists.

The main idea of *universals structuralism* is that mathematics is concerned with universals in the form of general patterns (or structures) that find their exemplification in concrete structured systems. The approach is developed in the works of Michael D. Resnik and Steward Shapiro. To quote from Resnik:

> [M]athematics is a science of patterns with mathematical objects being positions in patterns. ... The objects of mathematics ... are themselves atoms, structureless points, or positions in structures. As such they have no identity or distinguished features outside a structure. ... I take pattern to consist of one or more objects, which I call positions that stand in various relationships (Resnik, 1997, 199–203)

Shapiro, one of the most active contemporary proponents of the structural approach within the philosophy of mathematics, defines a system to be "a collection of objects with certain relations among them" Shapiro (2000, p. 259, 2011) and a structure to be "the abstract form of a system, which ignores or abstracts away from any features of the objects that do not bear on the relations" Shapiro (2011), "highlighting the interrelationships among the objects, and ignoring any features of them that do not affect how they relate to other objects in the system" (Shapiro, 2000, p. 259).

Shapiro maintains that such structures "exist objectively, independently of the mathematician, and independently of whether the structures have any exemplifications in the nonmathematical realm". That is why he calls his position *ante rem structuralism*: although there may be many possible concrete systems exemplifying any particular structure (as well as none at all), such a structure exists independently as "one-over-many" Platonic universal.

Shapiro's approach explicates an intuition, held by many mathematicians, that mathematical entities inhabit some sort of Platonic heaven. Only that for a structuralist these entities are not separate objects, but rather whole mathematical structures. They are supposed to be a sort of "pure structures" which only have the relevant structural properties while their instantiations, Shapiro's systems, always exhibit some specific, non-structural features.

In contrast to his own "ante rem structuralism," Shapiro (2014) also specifies another possible position which might be named *in rebus structuralism* (or "in re structuralism" as Shapiro himself does). It would be an Aristotelian view which would differ from the previous Platonic one in what makes structures exist. For an in rebus structuralist "the only structures that exist are those that are exemplified". Although the position is theoretically possible, Shapiro immediately adds that he is not aware of "any philosophers of mathematics who articulate such a view in detail".

Obviously, if one considers structures as a sort of universals, there is a whole multitude of different shades of universals structuralisms conceivable. Any general position vis-à-vis universals is simply mirrored and incorporated into mathematical structuralism. Ante rem (Platonic) structuralism and in rebus (Aristotelian) structuralism represent the two most obvious examples.

Although universals structuralism solves the problem model structuralism had with structurally irrelevant features of different models of the same structure, it also raises news problems of its own. The questions of the ontic status of universals and of our epistemic access to them have bemused philosophers for millennia. This is why Michael Dummett dubs this version of mathematical structuralism as "mystical" (Dummet, 1991, p. 295–6).

Those are the reasons why some philosophers chose paths different from the usual Platonic realism of universals structuralism. One of such paths we shall consider presently.

6 Modal structuralism

The main idea of Geoffrey Hellman's *modal structuralism* is to speak about structural features of possible systems (possible concrete structures). A modal structuralist is interested in studying what holds in imaginable concrete systems that share the same structure rather than invoking any pure (abstract) structures. Their arguments are of the true-in-any-model-of-the-given-structure sort. They also prefer not to quantify over "all systems of the same structure" or any such entities (cf. Hellman 1996, 2001, 2005).

As Shapiro (2014) explains "[o]n such a view, apparent talk of structures is only a *façon de parler*, a way of talking about systems that are structured in a certain way". Due to its hallmark which is to do without generic structures regarded as universals, this approach is also called "eliminative structuralism" or a "structuralism without structures". Let us sum up the main features of the modal structuralism as well as the motivation behind it in the words of its main proponent:

> Mathematics is the free exploration of structural possibilities, pursued by (more or less) rigorous means.
>
> [In a] modal-structural approach ... literal quantification over structures and mappings among them is eliminated in favor of sentences with modal operators. ... typical mathematical theorems are represented as modal universal conditionals asserting what would necessarily hold in any structure of the appropriate type that there might be.
>
> The first three approaches [model structuralism, category structuralism, universals structuralism] are framed in modal-free languages but they are entangled well above the neck (naturally) in Plato's beard. Sets, categories, or universals are just taken as part of reality, leading to perennial disputation as to the nature of such 'things', how we can have knowledge of them or refer to them, etc, ... Modal structuralism avoids commitment to such *abstracta*, at least in its initial stages ..., and raises the prospect that a (modal) nominalistic framework may suffice to represent the bulk of ordinary mathematics. ... The price of course is taking a logical modality as primitive, raising questions of evidence and epistemic access not unlike those raised by platonist ontologies. (Hellman, 1996, pp. 100–3)

The previous quotation sums up well the concerns that led to the formulation of this approach, its advantages over rival positions, as well as some of its limitations. Other drawbacks include its being rather cumbersome and, perhaps, feels less natural than the straightforward universals structuralism.

7 Category theory is implicit structuralism

The basic notions of category theory are *morphisms* (or simply arrows) and *objects* (or simply dots). Category is anythig that satisfies several basic axioms:

- Each pair of connecting arrows (f and g such that the first one ends in the same object the second one starts) have a unique composition arrow which starts where the first arrow does and ends with the second one (denoted as $g \circ f$ and read "g after f")

- Each object (A) is equipped with an identity arrow (1_A) which both starts and ends at the given object (A) and which works as both left and right identity of the arrows composition (i.e. $1_A \circ f = f$ and $g \circ 1_A = g$ for any convenient f and g)

- The composition of arrows is associative, that is, if we have three connecting arrows, it does not matter whether we first compose the first two and then compose the result with the last one, or whether we start composing from the last pair and compose the result with the first one (that is: $h \circ (g \circ f) = (h \circ g) \circ f$ for any f, g and h allowing such compositions)

According to the implicit structuralisms, mathematicians think about some axiomatically given structured systems, searching for regularities existing therein. Modern and epoch-making as category theory may be, it still is an exercise conducted by the means usual for the mathematical profession. Whether the entities of main interest are called sets or categories, it does not change anything on this general setup. Category theory surely is a clear example of implicit structuralism, and one of the axiomatic form too.

8 Category theory can be formalist structuralism

Although category theory, as is usually the case in mathematics, is mostly carried out informally, there are no barriers preventing it from being all undertaken formally. Category theory lends itself to formal structuralism with the same ease as any other part of modern mathematics does.

9 Category theory as universals structuralism

Category theory can also easily be considered as an example of the universals structuralism. The only novelty now is that the universals are the "worlds of structures" (i.e. categories) instead of the "structures" themselves. This is exactly why Hellman branded category theory as Platonist in the already quoted passage (Hellman, 1996, p. 103).

Any category is itself an entity structured by the network of its morphisms. Whether we draw it on a paper or imagine in our heads, we are dealing with a general structure. Mathematicians, many of whom are Platonists, naturally imagine such structures to subsist independently of our mind in some equivalent of Platonic heaven. If we imagine categories as Resnik's patterns or Shapiro's structures (I repeat: the whole categories, not their objects only), their universals structuralism seems to grasp category theoretic exercise quite accurately.

10 Category theory as modal structuralism

I do not see any reasons, apart from possible Hellman's lack of interest, why category theory could not receive the modal-structuralist treatment too. McLarty (2004, p. 45) explicitly considers this possibility, and finds a potential for "categorical nominalism" in this direction.

Category theory is thus either a straight example of, or is at least compatible with, four of our five different versions of mathematical structuralism. Yet, category theory, a foundational approach rival to set theory, surely cannot be classed as a model structuralism, which is usually connected with set theory. Or can it?

11 Category version of model structuralism

If we consider model structuralism in a more general way as taking some pre-supplied entities and adding some structure to them, or by means of them, it is exactly what is going on in category theory. Constructions in categories (terminal objects, products, limits in general, exponentials, ...) and constructions on categories (products of categories, functors, functor categories, ...) all proceed by introducing more structure into or onto categories. One moves from a standard category to a Cartesian closed category, for example, in the same general way as one moves from a set to an ordered set.

To put it differently, if we model structures by categories with some specific features, category theory can be recognized as an example of model structuralism. The only difference from the set-model variant is that structures are not "sets with extra structure" but "categories with extra structure".

Same as above, the structures are the whole categories here, not their objects only. I am well aware of the category theory practice to treat as structures object of (some) categories, not the categories themselves. Indeed, it is in this feature that I see the main contribution of category theory to mathematical structuralism. This contribution I am prepared to consider in a different paper. For the moment, categories themselves are structured mathematical entities – structured by their own morphisms, that is. Whether we call them structures or hyper-structures, the question is how they should be addressed by mathematical structuralism. And the answer is that it does not seem necessary to look for any new ways how to address them since the old ways of addressing structured mathematical entities seem to do just well. Which leads to the conclusion that, on the general level, category theory is not a new version of mathematical structuralism, but rather it subsumes well under any of the five more general versions considered.

12 Conclusion

As I have noted in the introduction, I am convinced of a high relevance of category theory for mathematical structuralism and I am prepared to defend this claim elsewhere. This relevance, however, does not consist in category theory providing another competing general philosophical framework for mathematical structuralism. After all, category theory is only mathematics. If philosophers of mathematics were correct in explaining mathematics, their explanations should cover category theory, too. I argue that they do. This might also be the reason why category theory (often) does not receive a special treatment by the philosophers of mathematics. If they believe that it fits naturally into one of the slots in their classification of structuralism already, there is no need to provide a new bracket.

References

[1] Awodey, S. (1996). "Structure in Mathematics and Logic: A Categorical Perspective". *Philosophia Mathematica*, **4**(3), 209–237. doi:10.1093/philmat/4.3.209.

[2] Awodey, S. (2004). "An Answer to Hellman's Question: 'Does Category Theory Provide a Framework for Mathematical Structuralism?'". *Philosophia Mathematica*, **12**(1), 54–64. doi:10.1093/philmat/12.1.54.

[3] Benacerraf, P. (1965). "What numbers could not be". In Benacerraf, P. and Putnam, H. (1983). *Philosophy of Mathematics; Selected readings*. Cambridge: Cambridge University Press. 272–294.

[4] Carter, J. (2005). "Individuation of objects – a problem for structuralism?". *Synthese*, **143**(3), 291–307. doi:10.1007/s11229-005-0848-x.

[5] Carter, J. (2007). "Structuralism as a philosophy of mathematical practice". *Synthese*, **163**(2), 119–131. doi:10.1007/s11229-007-9169-6.

[6] Cole, J. C. (2010). "Mathematical Structuralism Today". *Philosophy Compass*, **5**(8), 689–699. doi:10.1111/j.1747-9991.2010.00308.x.

[7] Dummett, M. (1991). *Frege: Philosophy of Mathematics*. Cambridge, MA: Harvard University Press, ISBN 0-674-31935-4.

[8] Hale, B. (1996). "Structuralism's Unpaid Epistemological Debts". *Philosophia Mathematica*, **4**(2), 124–147. doi:10.1093/philmat/4.2.124.

[9] Hellman, G. (1996). "Structuralism Without Structures". *Philosophia Mathematica*, **4**(2), 100–123. doi:10.1093/philmat/4.2.100.

[10] Hellman, G. (2001). "Three Varieties of Mathematical Structuralism". *Philosophia Mathematica*, **9**(2), 184–211. doi:10.1093/philmat/9.2.184.

[11] Hellman, G. (2003). "Does Category Theory Provide a Framework for Mathematical Structuralism?". *Philosophia Mathematica*, **11**(2), 129–157. doi:10.1093/philmat/11.2.129.

[12] Hellman, G. (2005). "Structuralism", In *The Oxford handbook of philosophy of mathematics and logic*. Oxford: Oxford University Press.

[13] Horsten, L. (2012). "Philosophy of Mathematics", *The Stanford Encyclopedia of Philosophy*. (Summer 2016 Edition), Edward N. Zalta (ed.), https://plato.stanford.edu/archives/win2016/entries/philosophy-mathematics/, [retrieved November 14, 2016].

[14] Landry, E. (2009). "How to be a structuralist all the way down". *Synthese*, **179**(3), 435–454. doi:10.1007/s11229-009-9691-9.

[15] Landry, E. (2016). "Mathematical Structuralism". *Oxford Bibliographies Online Datasets*. doi:10.1093/obo/9780195396577-0305 [retrieved November 14, 2016].

[16] Mac Lane, S. (1996). "Structure in Mathematics". *Philosophia Mathematica*, **4**(2), 174–183. doi:10.1093/philmat/4.2.174.

[17] Mac Lane, S. M. (1997). *Categories for the working mathematician*. 2nd ed., New York: Springer, ISBN-10: 0387984038.

[18] McLarty, C. (2004). "Exploring Categorical Structuralism". *Philosophia Mathematica*, **12**(1), 37–53. doi:10.1093/philmat/12.1.37.

[19] Reck, E. H. and Price, M. P. (2000). "Structures and structuralism in Contemporary Philosophy of Mathematics". *Synthese*, **125**(3), 341–383. doi: 10.1023/a:1005203923553.

[20] Reck, E. H. (2003). "Dedekind's Structuralism: An Interpretation and Partial Defense". *Synthese*, **137**(3), 369–419. doi:10.1023/b:synt.0000004903.11236.91.

[21] Resnik, M. D. (1997). *Mathematics as a science of patterns*. Oxford: Clarendon Press, ISBN-10: 0198250142.

[22] Shapiro, S. (1997). *Philosophy of mathematics: Structure and ontology*. Oxford: Oxford University Press, ISBN-10: 0195139305.

[23] Shapiro, S. (2000). *Thinking about mathematics: The philosophy of mathematics*. Oxford: Oxford University Press, ISBN-13: 9780192893062.

[24] Shapiro, S. (2005). "Categories, Structures, and the Frege-Hilbert Controversy: The Status of Meta-mathematics". *Philosophia Mathematica*, **13**(1), 61–77. doi:10.1093/philmat/nki007.

[25] Shapiro, S. (2011). "Structuralism in the philosophy of mathematics". *Routledge Encyclopedia of Philosophy*, Taylor and Francis, https://www.rep.routledge.com/articles/thematic/structuralism-in-the-philosophy-of-mathematics/v-1, doi:10.4324/9780415249126-Y095-1, [retrieved November 14, 2016].

[26] Shapiro, S. (2014). "Mathematical Structuralism". *The Internet Encyclopedia of Philosophy*, ISSN 2161-0002, http://www.iep.utm.edu/, [retrieved November 14, 2016].

[27] Weaver, G. (1998). "Structuralism and Representation Theorems". *Philosophia Mathematica*, **6**(3), 257–271. doi:10.1093/philmat/6.3.257.

www.ingramcontent.com/pod-product-compliance
Lightning Source LLC
Chambersburg PA
CBHW080439110426
42743CB00016B/3212